Abdominal Fitness

By

Mohamed F. El-Hewie

TABLE OF CONTENTS

CHAPTER 1: THE HUMAN BELLY

1.1. Man-made foods

Humans differ from the rest of the animal species in their eating habits in such manner that distinguishes the human belly among the animal kingdom. We are the only animal species that cooks food and alters the chemical and physical properties of organic matters for dietary consumption. We make **cheese** and **butter** from milk, **sugar** from fruits, **oil** and **margarine** from plant seeds, and **salt** from the sea water. The man-made **high fat contents** of cheese, butter, oil, and margarine do no good to our biological system since humans have industrialized societies that require minimal physical activities, inadequate to burn the high fat intake. Aside from their high caloric contents, those man-made fats require greater production of emulsifying enzymes and salts in order to break down their molecules into absorbable substances. On the other hand, sugar and salt are highly soluble substances that increase the **osmotic pressure** of the blood leading to the net of leaking fluids from the blood capillaries into the intercellular space. In complete contrast to man-made food, proteins are too complex to be made by man on mass production scale. Further, proteins are easier to digest without emulsification and have the greatest **oncotic pressure** in the blood stream due to the colloidal nature of protein molecules. In simple words, the large size of the protein molecules, compared to the molecules of sugar, salt, and fat, causes the protein molecules to pull fluids into the blood stream and from the extracellular space.

With only those six, man-made food ingredients, millions of people are afflicted by many life-long diseases that could have easily been prevented by the mere avoidance of those foods. The human belly is fattened, enlarged, and sluggish by the wide spread consumption of cheese, butter, oil and margarine. Those four food ingredients alone challenge the ability of the digestive system in emulsifying fat, far and beyond the natural limits of dealing with greasy substances, by the pancreas, stomach, liver, and intestine. The **deceleration of the intestinal motility** due to the consumption of those food leads to indigestion, constipation, and distension, which is nothing short of partial intestinal blockage that many people take for granted as normal way of life.

After the brutal digestion, emulsification, and absorption of the four man-made fatty foods, our body has to isolate them in remote stores; under the skin, within the surrounding of the internal organs, and anywhere else that does not impeded the natural flow of fluids in the body. In other words, the extra fat consumption is handled by the body with seclusion, not utilization or beneficence. Before the process of seclusion of consumed fats ends by the sudden death of the patient with coronary artery diseases, the belly issues many warnings along two or three decades, before **heart attack** shows its ugly face. The belly enlargement, fattening, and discomfort, along those decades of life, should alert the individual of the ongoing alteration of his/her body by the insulting food ingredient.

The remaining two; sugar and salt, are the major culprits is disturbing the homeostasis of the body. The molecules of sugar and salt possess the highest osmotic pressures that enable those molecules to sneak from the blood stream into the space surrounding cells within minutes of their consumption. Thus, salt and sugar act as loose, unrestrained trespassers on the extracellular space, where neither the heart nor the kidney has access to. Swelling ensues with the accumulation of fluids in the extracellular space. As the blood volume loses its water content to the extracellular space with the excessive consumption of sugar and salt, the nervous system reacts by raising the **blood pressure** in order to compensate for the diminished volume of the blood. The high blood pressure wears down the blood capillaries and the arterioles in a similar fashion to the high vehicular traffic on a highway road. The blood capillaries of the heart and kidneys give way to the high blood pressure by rupturing causing **microscopic blood leakages**. If the blood pressure is not restored to its normal level, the number of the microscopic hemorrhages reach the level that causes gross scarring and destruction of the tissue of the heart and the kidneys. High blood pressure also traumatize the walls of large blood vessels in the same fashion of causing microscopic hemorrhages. Those heal by trapping cholesterol substances that **hardens the wall** of the blood vessels. The hardened blood vessels steal the joy out of the life of the individual by rendering the individual incapable of living active lifestyle.

****

With only those six, man-made food ingredients, millions of people are afflicted by many life-long diseases that could have easily been prevented by the mere avoidance of those foods.

****

1.2. Role of culture on eating habits

Second, humans are highly sociable creatures motivated by emotion in excess of reason. As such, the family environment, peer pressure, and community advertisement play great role in determining the eating habits of people. Observing the **role of culture** on the waistline of people, one could easily read into the eating habits of different cultures. In an article entitled "Waist Circumference Percentiles in Children and Adolescents", in Current GGH - Vol. 24, March 2005, the authors acknowledged the relationship between the waist-hip ratios in adults and children with the development of **hyperlipidemia, insulin resistance, hypertension,** and **diabetes mellitus**. The article speculated that waist circumference alone may be a more useful and more easily obtainable index in both adults and children. On the cultural differences, the article concluded that Mexican-American boys and girls have higher waist circumferences than African-American or European-American children at each age, while African-American boys have lower circumferences than the other ethnic groups. African-American boys have a slower rate of increase in waist circumference as they age than do the other boys. Mexican-American girls have the fastest rate of increase of all girls. At the 75th percentile, 16 and 17 year old Mexican-American and African-American girls exceed the waist circumference **cut-off** point for obesity related co-morbidities in adult women.

****

The relationship between the waist-hip ratios in adults and children with the development of hyperlipidemia, insulin resistance, hypertension, and diabetes mellitus.

****

1.3. Modernity and abdominal illness

Third, the human belly has been impacted by the fast **development of civilization** in the last few centuries. The discovery of electricity afforded great portion of the population the opportunity to stay late, work and entertain during the overnight hours, with the devastating alteration of the biological processes of the body. With the long hours of awakening, people consume greater amounts of food and exercise less, thus develop abdominal illnesses. Further, the invention of freezers, microwave ovens, and other kitchen ware, has contributed to the greater consumption of and access to food. Thus, individual discretion is the major determinant in all maladies that afflict the belly, which in turn, afflict the general health of the individual, both on the short term as well as the long.

CHAPTER 2: PUMPS AND FILTERS

2.1. The heart and the kidneys

In order to live, animals must filter their blood from gases generated by the metabolic processes of the cells of their body. The filtration process is called "**gas exchange**" and takes place in the **lungs**. In aquatic animals, gills exchange gases. The metabolic gas products are exchanged between the blood and the inspired air such that oxygen enters into the blood stream and carbon dioxide departs to the ambient air. The air carries the gases required to burn the energy in food. All foods consumed by animals must be converted into fluids in order to be processed through digestion, absorption, and metabolism. Animals are equipped with **kidneys** that filter fluids such that the net water content of the animal remains within the limits essential for survival.

The lungs filter air at extremely high rates compared to the rates of filtering fluids by the kidneys. For example, a healthy person breathes between thirteen to sixteen times every minute, yet urinates only few times during the entire day. With such dual filtering techniques, of **slow fluid filtration** and **fast air exchange**, animals manage to deal with variation in mobility by increasing the rate and depth of breathing in response to the demand for energy by muscles and vital cells. Vigorous acts such as coughing, laughing, bearing down, yelling, running, etc, increase the breathing rate and depth according to the intensity of the action.

The dual filtration system of the lungs and kidneys is interdependent. As the two filters manage the ingredients of the body fluids and gases, the lungs could filter molecules which were originally ingested as fluids. Likewise, the kidneys could filter, molecules which were originally breathed as gases. such interchange of molecules between food and breathed air results in controlling the **acidity** and **alkalinity** of the body fluids. The simple explanation for such scientific jargon is that chemical molecules react with each other according to their characteristic potential to **give** or **take** electrons into their electronic shells. The electronic composition of molecules is the dominant chemical language by which molecules communicate.

Therefore, one could clearly discern that the lungs and kidneys are proper molecular filters that manage the chemical milieu of the body fluids. The body homeostasis, which is another name for the body internal environment, is thus controlled by the two **macro filters**; the lungs and kidneys. In between the two, there are trillions of **micro-filters** in the forms of cell membranes, nuclear membranes, and other membranes of inter-cellular granules and structures. Both systems of filters communicate and integrate through sophisticated chemical, neural, and hormonal systems aimed at maintaining the body internal environment optimum for living.

Filters require flow of molecules in order to sort out the streams of traffic between the inside and the outside of the body. The **heart** is the primary pump that automates the flow of fluids around the body. The heart pump is unique in its automaticity of generating electrical signals even when the heart is removed from the body of the animal. Within the body, the heart plays by the rules of the nervous system by responding to slowing and speeding signals sent to it via the **vagus nerve**. The heart pumps blood to the kidneys and lungs as an essential function to filtration of fluids and gases. It is noteworthy to mention that the heart must initially pump blood to the lungs before the remaining of the body. The heart-lung pathway is called the **cardio-pulmonary** circulation. Subsequent to that pathway, the heart pumps blood to the entire system via the **systemic cardiac circulation**. In the latter, the kidneys receive its share of fluids to filter.

2.2. The voluntary abdominal pump

Apart from the automatic and enduring heart pump, the **belly** or abdomen functions as the voluntary pump controlled by the will and emotion of the animal. The abdominal muscles could not automate electric production of signals and must be contracted by the individual in order to perform their task. In contrast to the heart pump, the belly pumps both gases and fluids as follows.

1. The belly performs the formidable task of **sucking the blood return** from the legs, genitalias, and guts and forwarding such blood flow to the heart. The suction action of the belly is initiated by the elevation of the **diaphragm** and relaxation of the **pelvic floor** of the belly. The diaphragm comprises the roof of the belly. It rises up as the chest wall expands during inspiration. The ribs elevate to increase the size of the chest cage. The rising ribs pull the diaphragm upwards and outwards, thus expanding the size of the belly, as well. The pelvic muscles comprise the floor of the belly, which give way in order to increase the size of the abdominal sac. Thus the sac of the belly acts as an expanding balloon that sucks blood from the lower body. The belly balloon then tightens its floor and lowers its roof, except that the diaphragm has openings that permit the inferior vena cava (the major vain that transport blood from the lower body to the heart), the esophagus (the tube that connects the mouth with the stomach), and the major blood and lymph vessels, to traverse between the belly and the chest.

2. The belly assists in ejecting semen during ejaculation by the combined actions of the four walls (considering the spinal erectors as the fourth wall, in addition to the abdominal sac, the pelvic muscles, and the diaphragm) . The diaphragm and floor of the belly are supported by the **abdominal muscles sacs** in squeezing the pelvic contents such that both penile erection and vesicular contraction could sustain the flow of semen and the hardening of the penis. In females, erection, congestion, contraction, and lubrication are induced and maintained by the same muscular actions. The fourth and always overlooked muscle group, beside the floor, diaphragm, and abdominal muscle sac, is the **spinal erector**. The spinal erectors support the spinal discs from the back of the body during the extreme pressure within the belly, during ejaculation, defecation, bearing down, screaming, coughing, or laughing.

3. The belly coordinates breathing during vigorous activities by expanding the lung cavities by relaxing its floor and abdominal muscle sac during deep inhalation. During vigorous expiration, not only the three muscles walls of the belly must contract, but also the spinal erectors must antagonize the abdominal pressure in order to avert spinal disc bulging or herniation. Abdominal breathing dominates aggressive activities where the belly acts as **extension reservoir** attached to the floor of the lungs. In young children, abdominal breathing dominates during normal activities due to the underdeveloped bones of young children. Growing up causes the pelvic bones to expand and harden such that the guts fall into the pelvic cavity, instead of bulging in the front of the body.

4. The belly regulates the blood flow within the **spinal cord**, muscles, and other structures, through the rhythmic alternation in abdominal pressure. That occurs during Valsalva's maneuver, for example, as well as throughout all abdominal contractions. When the abdominal muscles thin out, weakens, and lose their robust tone, the spinal

circulation gets sluggish, spinal muscles atrophy or weaken, spinal bones become fragile, and spinal ligaments detach, tear, or slow to repair.

5. The belly mobilizes the intestines and **intestinal contents** and transmits the intestinal pressure onto other organs. The rhythmic contraction of the four walls of the belly (considering the spinal erector as a back wall) maintain the flow of food from the mouth to the rectum. weak or unsupported belly cause the intestinal contents to press against weak points and thus bulge into hernias.

6. The belly mobilizes the extravascular fluids such that cells could gain access to new nutrients after spoiling their immediate surrounding fluid environment. The red blood crosses from the arteries through the **cellular neighborhood**, it returns as blue blood, through the veins with the help of the abdominal rhythmic contraction. That is simply a matter of inducing pressure change of the red blood in order to cause its accelerated crossing to the veins, while nourishing the cells.

****

Apart from the automatic and enduring heart pump, the belly or abdomen functions as the voluntary pump controlled by the will and emotion of the animal.

****

CHAPTER 3: INSIDE THE GUTS

3.1. Intestinal contents

The **intestinal contents** could be most intriguing to people who never looked inside an open incision of a human abdomen. Looking inside a skinned turkey or chicken, somehow does not get people to make the connection between our guts and those of other fellow animals. The more interesting fact is that most people could not grasp the direct cause between oral intake and belly fat. Even in the most technically experienced people, I found out that many could not relate the simple example of filling gas tank through the intake opening of such gas tank. Filling a tank with gas through a small opening is like fattening the belly with fat through oral intake. Fat does not deposit on the belly from the air! Such an obvious statement is not always obvious for others.

Even medical students use mnemonics such as 5F to remember the intestinal contents. Those are: **fluid, feces, fat, fetus,** and/or **flatus.** With the exception of the fetus, the remaining intestinal contents relate to the pumping function of the belly. The belly maintains the flow of fluids from the mouth to the rectum with great ability to absorb special nutritional gradients before the intestine could dry out the adequately formed stool for final elimination. Gases accumulate in the intestine when the individual is sedentary and unable to maintain adequate level of mobility so as to reduce the putrefaction or decay of the remains of the undigested food. finally, excess energy intake is stored as fat within the abdominal organs and under the skin, until the individual curtails his/her intake of calories so as to deplete the stored fat energy.

Fluid accumulation occurs in disease conditions that affect either the poor pumping of fluids by the **heart** or the poor removal of fluids by the **kidneys.** Fluids doe not accumulate within the intestines since those has great ability to absorb and remove fluid from food. The small intestine maintains the fluid content of the food until it is mostly digested and its absorbable components are taken out by the body. That process might take up to nine hours after mean. The large intestine is the main fluid absorbing portion of the intestine. If the large intestine fails to absorb fluids due to inflammation or due to osmotic medication, diarrhea ensues. So, where does the fluid accumulate? In the cavity between the intestine and the abdominal wall. That is called **"potential"** cavity because it collapses on its walls if there were not fluids to fill it. When the flow of fluid faces low pressure change, it stagnates. That occurs when the heart fails or the kidneys malfunction. In malnourished kids, the low protein intake causes fluids to leak from the capillaries into the outer vascular space.

3.2. Uterine contents

In pregnancy, the **fetus** is added to the normal intestinal contents. As pregnancy develops, the enlarging fetus compresses the kidneys, ureters (the urine ducts between the kidneys and bladder), and the inferior vena cava and thus causes accumulation of fluid, in addition to the fetal mass. If the fetus compresses the kidneys severe enough to cause high blood pressure, a condition called **"preeclampia"** ensues. The high blood pressure, high protein content in urine, and accumulation of fluids in a pregnant women required immediate medical attention in order to save the carrier and the passenger. In its extreme, preeclampia progresses to **eclampsia** when seizure develop. This is an emergent medical condition.

Figure 1. The fetus is not the only addition to the pregnant belly. Fluids could accumulate in the belly before the legs swell due to the increased intraabdominal contents. Note that the **pregnant abdomen** does not descend or shadow the genitalia as is the case in extreme fat deposition. Here, the Great Omentum is not thickened, but rather, its is the uterus that is filled with fetus and placenta.

3.3. Fat depots
Fat has unique spots for deposition beside those observed from outside. The greatest reservoir of fat is the **Great Omentum.** That is a fold of the inner lining of the abdominal sac that hangs over and in front of the stomach and descends under the belly button. In extreme conditions of massive fat deposition, the Great Omentum could carry more fat then the entire weight of the rest of the body. Fat also accumulate within the muscles of the abdominal sac and intermingles with the muscle fibers. Fat could also accumulate under the skin in massive amounts.

Figure 2. Fat deposits in the Great Omentum before it does under the skin. Note the difference between fat belly and pregnant belly. Here, the thick and fat Great Omentum, in addition to the excessive fat accumulation **under the skin**, lead to the descent of the belly over the genitalia. The fattened abdominal sac could not mobilize the intestines and thus causes intestinal distension. The distended intestine impairs the breathing and interferes with exercise. The lack of exercise cause more fat deposition. The vicious circuit continues unless intake in curtailed.

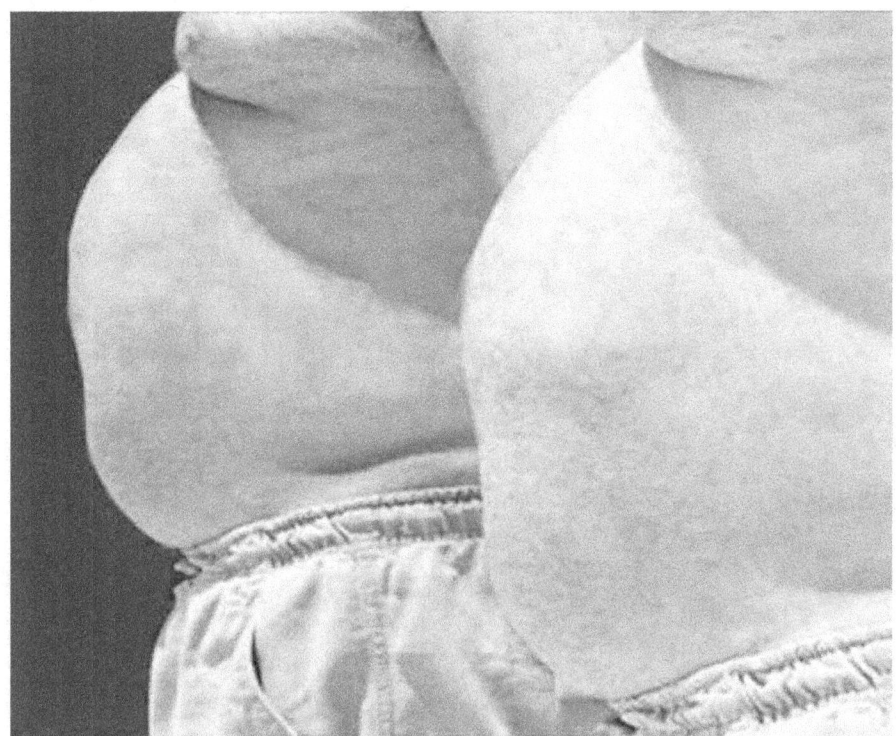

Figure 3. The excessive deposition of fat in the Great Omentum and the distention of the colon cause such inflammation of the skin due to poor circulation and poor hygiene.

****

The greatest reservoir of fat is the Great Omentum. That is a fold of the inner lining of the abdominal sac that hangs over and in front of the stomach and descends under the belly button. In extreme conditions of massive fat deposition, the Great Omentum could carry more fat then the entire weight of the rest of the body.
****

3.4. Early signs of belly chaos

The neglect of belly hygiene is life threatening in mentally retarded patients and causes large number of intestinal obstruction and rupture. The inability of the individual to exert adequate discretion in regulating the quantity and type of food that ensures healthy body leads to many losses of lives due to extreme overweight, loss of ability to earn living and to maintain proper sanitary conditions.

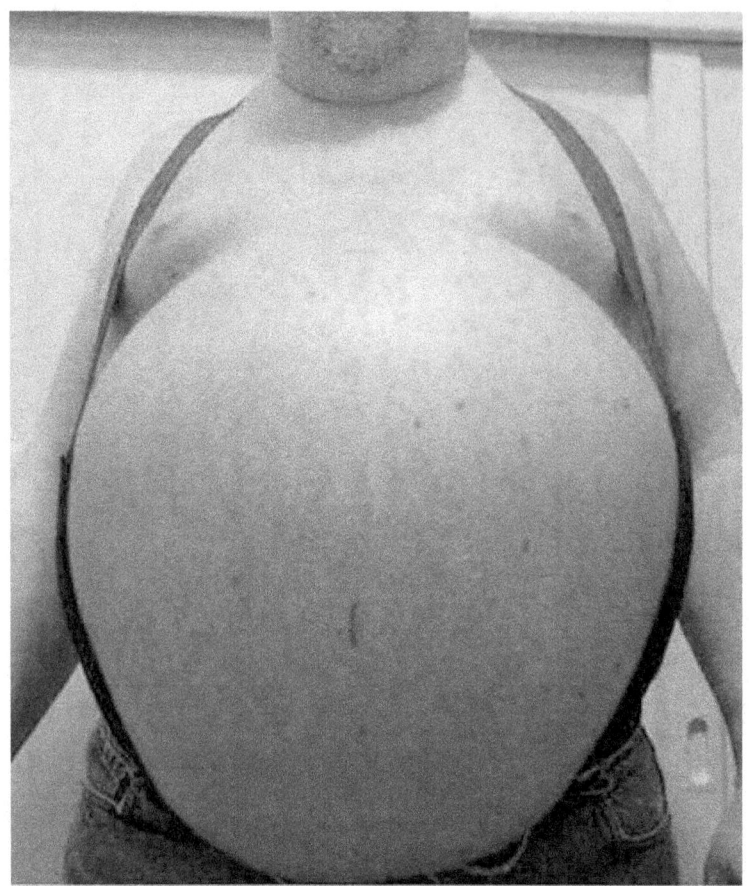

Figure 4. Extreme abdominal distention has many underlying causes than simple constipation. Alcohol consumption in addition to excessive intake of fat will get the individual to such extreme case of hugely enlarged colons. Here, the over pressurized belly acts as a tourniquet on the major blood vessels and thus causes extreme pains in the limbs and spine. The patient is entirely incapacitated with his condition and must be assisted with medical interference in order to regain his ability to breathe. move, and rest. This is a patient is agonizing pain, not just disfiguring colon enlargement.

Flatus and **feces** become burdensome when mobility decrease, fattening food is consumed in excess, of the person lacks common knowledge of how to monitor simple symptoms of troubled belly. Accumulation of feces and flatus could initiate the above scenario of extreme fat deposition that is notorious to reverse. Change of odor or convenience to eliminate intestinal contents should alter the person that an **insult** is underway. Insults to the belly hygiene are numerous and identifiable. Food that causes constipation or delays elimination beyond 24 hours should be the first suspect. The **quantity of food** intake should be checked first in order to restore the intestinal ability to combat the ensuing insult. Food that changes the odor the stool or flatus is next in the exclusion process after the quantity of food has been minimized. It must be observed that restoring regular intestinal hygiene might require few days of exclusion of the insulting food. It is counter productive to try to add fibers that roughens the stools when the quantity of food intake is not reduced. Adequate intake of water is crucial to the softening process of intestinal contents. It should be duly stated that fluids that have any **caloric contents** could worsen abdominal distention when used to alleviate the hardening of stools.

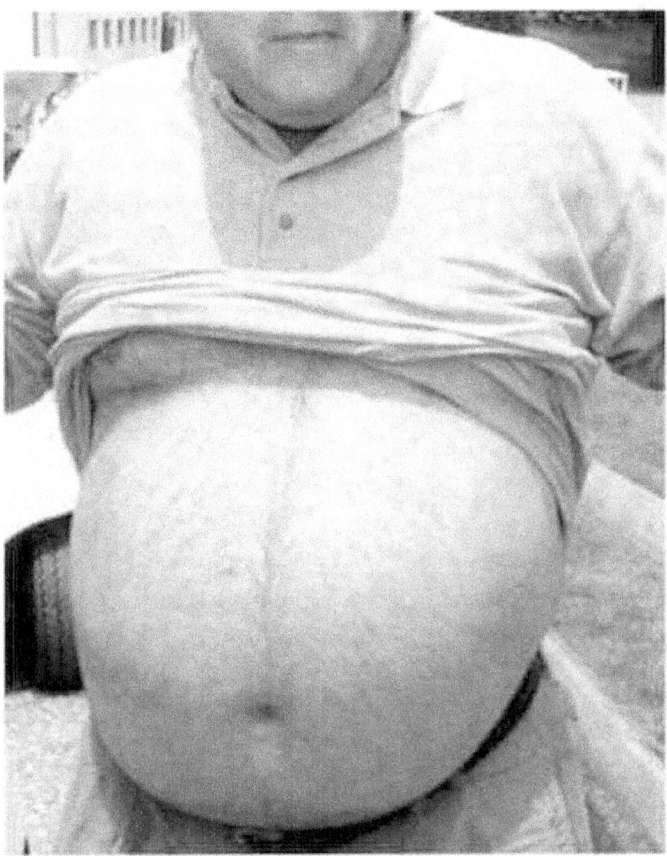

Figure 5. The accumulation of feces and flatus starts the process of progressive fat deposition due to the weakening of the abdominal muscles and the strain on the vital organs. Here, despite the slight fat deposition under the skin compared to **Figure 2**, the Great Omentum is the main reservoir of fat and lies under the middle line of the belly. The rounding of the belly is due to intestinal distension. Note the difference between a pregnant belly where the belly button points upwards and this fat belly where the belly button falls in the lower third of the abdomen. The pregnant belly in Figure 1, the fetus starts from the uterus and progresses upwards. In this fat belly, the Great Omentum accumulates fat from the top of the stomach and descends downwards.

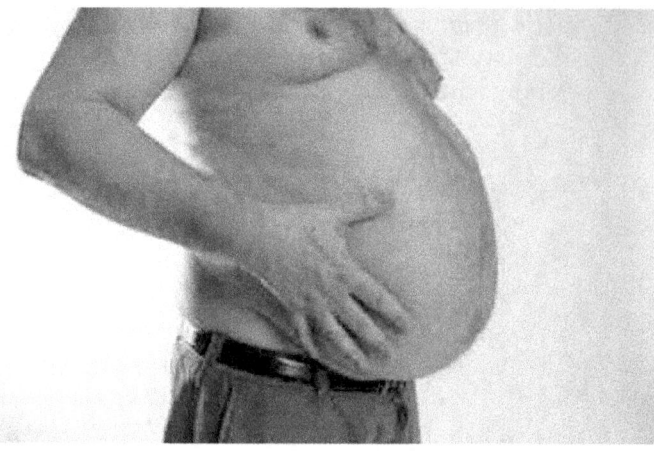

Figure 6. The vicious circuit of weakening of the abdominal muscles, accumulation of flatus and feces in the intestine, deposition of fat within the internal organs, and the thinning of muscle mass of the lower body. As the belly distends, the lower body lacks proper blood flow or exercise. The buttocks are wasted and the belly descends over the genitalia. The patient puts his hands of the stressed and distended colons that causes his back pain and inability to breathe properly during exertion.

****

It must be observed that restoring regular intestinal hygiene might require few days of exclusion of the insulting food. It is counter productive to try to add fibers that roughen the stools when the quantity of food intake is not reduced. Adequate intake of water is crucial to the softening process of intestinal contents. It should be duly stated that fluids that have any caloric contents could worsen abdominal distention when used to alleviate the hardening of stools.
****

3.5. Role of individual discretion

Figure 7. Getting ripped belly is a matter of eating discretion rather than exercise alone. The six packs of abdominal rectus serration are not due to excessive exercise, but rather low fat deposition under the skin and thinned out Greater Omentum. In such slim belly situation, the individual could maintain regular elimination, consistent exercise, and convenient breathing under extremes of stress and exertion.

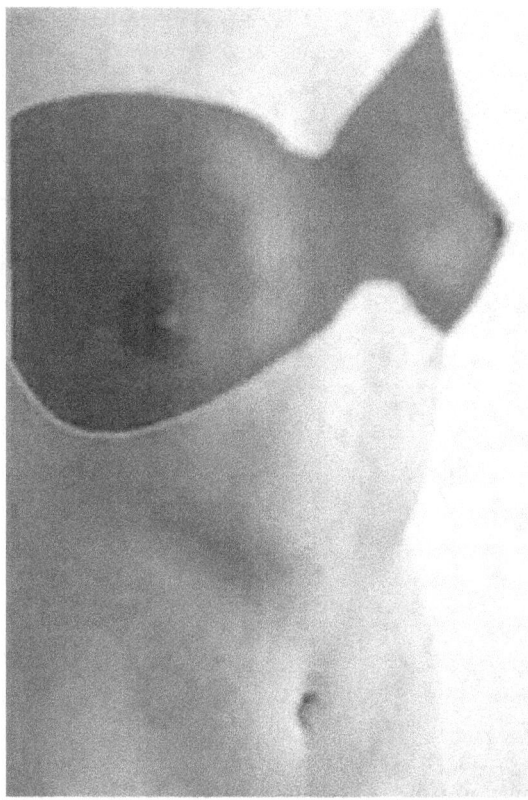

Figure 8. Women struggle with developing ripped abdominal muscles due to the actions of estrogen in depositing fat under the skin. Due to the light bodyweight of women, estrogen's fat deposition protects women from bone thinning due to low absorption of calcium. Indeed, obese women are less subjected to osteoporosis than thin women. In this figure, the **abdminis external oblique** muscles pulls the belly towards the spine from both sides of the **abdominis rectus muscle**. The latter is the long muscles that form the two banks on each side of the middle line, passing by the belly button.

****

Getting ripped belly is a matter of eating discretion rather than exercise alone.

****

CHAPTER 4: INSIDE THE BELLY

4.1. The abdominal organs

The human abdomen contains most of the vital organs beside the heart, which is contained in the chest cavity and the brain which is located in the skull. The belly contains the **stomach, pancreas, liver, spleen, kidneys, adrenals, and small and large intestines, and the major blood vessels** (the aorta and inferior vena cava). The spinal cord bypasses the belly by traveling within the spinal canal in the vertebral column. Thus, the spinal cord escapes the direct pressure changes within the abdominal cavity. As such, the belly hosts three different hydrodynamic chemical systems in its cavity. (1) The **bacterially rich flora** of the intestine breaks down the food intake into its basic ingredients suitable for absorption from the intestine to the blood stream. (2) The **bacterial-free urine** clearing system of the kidneys that adjusts and maintains the molecular balance of the body fluids. (3) The oxygen and nutrient **blood delivery** system that uses the highways of the major blood vessels to deliver nutrients and remove waste.

4.2. The septic digestive system

Belly hygiene affects the three vital systems. The **bacterially rich intestinal flora** constitutes the major concern of belly hygiene. The other two systems are mostly intrinsic, automated, and require minimal individual discretion beyond that needed to stay mobile. Mainly, the quantity and quality of food intake will impact the intestinal system first, before that affects the other two systems. The cascade of events ensues by **impaired decay** of digested food, due to either poor quality of food or low level of mobility that caused the poor mixing of intestinal juices with the digested food. The improperly mixed food offends the bacterial distribution and leads to overgrowth of abnormal bacterial population. Hence, flatus accumulates, causes distention of the intestine and compression of the surrounding vital organs, and encroaching over the renal and vascular systems.

If the individual could intercept the initial insult of **skewed decay** of intestinal food remains, that could terminate the cycle of flatus accumulation, constipation, weakening of the abdominal muscles, and distress of the surrounding vital organs. However, society plays major role in guiding individuals to the proper course of action in such vital issues as belly hygiene. The chemical magic for the solution of health problems is overrated and dominant despite the clear role of elimination of the underlying causative insult. Mobility alone restore intestinal flora when then the quantity of food is moderated. This simple role, of increasing the mixing of intestinal contents with the intestinal juices by mobility and reducing those contents to cutting down on food intake, suffices for the intestinal bacterial flora to restore its original population and for the intestinal wall to heal its structural integrity.

If digestion continues to decline, flatus accumulates and the intestine distends. The **large colons** are the major putrefaction sites in the in the body. On the left, the large colon ascends as high as the bottom of the spleen. The large colon rests on the back wall of the abdomen and thus affects the spinal circulation of blood. A distended large colon causes pain in the **lower back** region due to congestion of the blood flow between the spinal veins and the abdominal vena cava. In short, the distended colon prevents the lower back muscles, tendons, and soft tissues from draining their blood into the abdominal veins. The person feels the bulging of the lower back against the internal bones, when trying to release the intestinal flatus. If the abdominal muscles were robust, those would have prevented the accumulated intestinal gasses for reaching such contained volume that impairs the draining of the spinal venous blood.

****

Mobility alone restore intestinal flora when then the quantity of food is moderated. This simple role, of increasing the mixing of intestinal contents with the intestinal juices by mobility and reducing those contents to cutting down on food intake, suffices for the intestinal bacterial flora to restore its original population and for the intestinal wall to heal its structural integrity.

****

4.2.1. Low back pain

Spinal congestion due to accumulated intestinal gases affects organs that derive their nerve supply from the lower back. Those are the pelvic organs and lower limbs. If the process of intestinal distention lasts for few years, both the lower limbs will suffer loss of muscular mass and sensation and the pelvic organs, such as the genitals, bladder, and rectum will follow to malfunction and atrophy. The direct remedy to spinal congestion of the lower back, in such scenario, is the relief of the intestinal destination or exercising the lower back such that the spinal erectors muscles could augment the abdominal muscle sac in reducing intestinal distention.

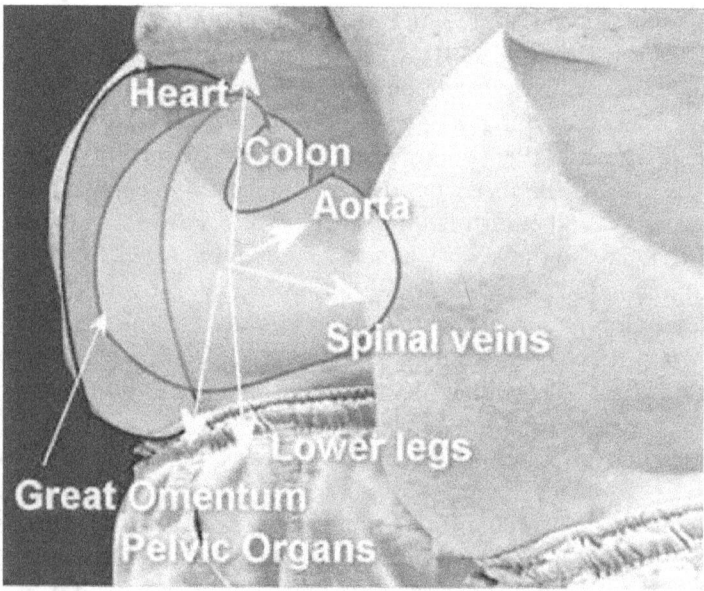

Figure 9. Poor belly hygiene in these two men affected the circulation between the heart and the spine and the lower limbs. In addition to skin ischemia (lack of blood supply), the lower limbs and the genitalia atrophy (tissue wastes away).

20

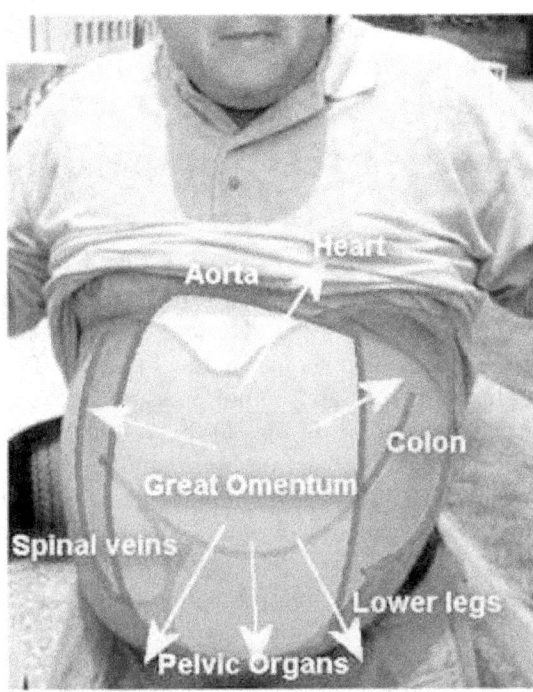

Figure 10. Even though the abdominal skin in this man appears in better hygienic condition than the previous ones, the abdominal distension and heavy Omentum are aiming at his lower legs, spine, and heart.

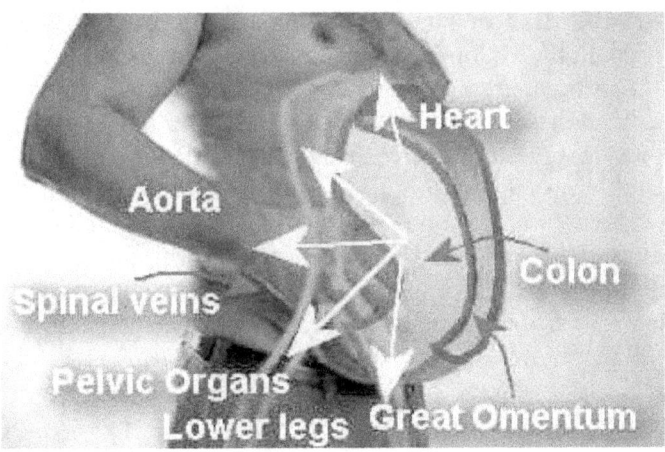

Figure 11. Even though this main is more concerned about his belly, his real troubles lie beyond that as his lower back is strained, his lower legs and genitalia are deprived of circulation, and his kidneys and heart are stressed with impaired breathing and abdominal pressure.

****

The direct remedy to spinal congestion of the lower back, in such scenario, is the relief of the intestinal destination or exercising the lower back such that the spinal erectors muscles could augment the abdominal muscle sac in reducing intestinal distention.

****

4.2.2. Compression of the kidneys

Poor belly hygiene impacts the general health of the individual beyond the local pain of the spine and impaired function of the lower limbs and genitalia. Those local maladies are dwarfed by the greater impact of the abdominal distension on the kidneys. The congestion of the kidney tissues, due to the compression of abdominal pressure over the major abdominal blood vessels, causes the kidneys to trigger the rise in the system blood pressure by releasing the **renin hormone**. Such paradoxical reaction by the kidneys aims to create greater blood pressure that could oppose the greater abdominal pressure. The kidneys are the only organs in the body that possess the hormonal regulation of the blood pressure during the low flow of blood through the kidneys. As such, a simple condition of poor belly hygiene that ensues due to inadequate mixing of food with the digestive intestinal juices, leads to such system reaction of increased blood pressure. This would require medication which in turns worsens the vicious circuit of intestinal distension and enlargement unless discreet reduction in the quantity of food intake in exercised.

4.2.3. Hernias or least resistance bulging

The sluggish and distended intestines drains the energy storage of the abdominal muscle sac, which gives away to weakness, thinness, and fat accumulation. A sedentary person develops plenty of fat zones within the flesh of his/her abdominal muscles. Those are mechanically inactive zones that weaken the abdominal muscles in opposition to the high pressure of the distended intestine. During aggressive daily chores, the intestinal pressure hikes to greater levels and thus exceeds the ability of the abdominal sac to contain its internal constituents. The imbalance in pressure, between the intestine and the under-toned abdominal muscles, leads to the **passage of the abdominal contents** through the spots of least resistance. Those are the natural openings through which vessels and tracts enter and exit the abdominal cavity. Among those least resistance passages are the inguinal canal; where the testicles pass from the belly to the scrotum, the femoral foramen; where the blood vessels cross from the belly to the lower legs, the obturator foramen; blood vessels pass from the pelvis to the lower legs; the openings of the diaphragm; where vessels and tract cross between the chest and the belly. That is beside the passages through newly created spots through the weakened and thinned out abdominal muscle sac.

Even though the above described intestinal hernias could be trapped and strangulated, thus leading to surgical emergencies, yet **intestinal twisting** and strangulation, due to the same process of pressure imbalance, could lead to acute abdomen and quick death.

4.3. The aseptic urinary system

The belly hosts the two kidneys in its back portion, under the lower ribs of the chest. The kidneys are the main regulator of **blood pressure** in response to fluid clearance. The kidney system is free from bacteria since it performs the most delicate function of sorting of molecules and clearing the excess through the urine. The kidneys also control the flow of calcium between the blood, bones, and urine. It does that by making **vitamin D** which regulates the deposition of calcium in the bones and by regulating the clearance of the excess phosphates from the blood, and into the urine.

The presence of the kidneys within the abdominal cavity subjects the kidneys to the disturbances in the septic digestive system. When digestion is hampered, the large and distended colon compresses both the kidneys and their blood vessels and urine tracts such that both flows of blood; to and out of the kidneys, are impeded by the pressurized colon, and the urine flow is blocked, for the same reason. The kidneys could endure the **intestinal disturbances** for many decades due to its millions of glomeruli that clear the urine from the blood. However, gradually and surely, the kidneys scars, atrophy, and shrinks when the mechanical pressure by intestinal contents protracts for long time.

The kidneys are directly attached to the suprarenal glands; the **adrenals**, that secrete four categories of hormones which regulate blood pressure, urinary clearance of minerals, sexual function, and sympathetic activity. Those hormones are as follows:
* glucocorticoids (e.g., cortisol): regulate blood pressure and resistance to infection.
* mineralocorticoids (e.g., aldosterone): regulate the mineral content of the body.
* androgens (e.g., testosterone, estrogen, progesterone): regulate the sexual characteristics of the individual.
* adrenaline (also called epinephrine) and noradrenaline (also called norepinephrine): regulate reaction to stress and emotion.

When the glands produce more or less hormones than required by the body, disease conditions may occur. Three categories of the adrenal hormones emerge from a single **cholesterol** compound, by the actions of cascades of enzymatic chains. When one enzymatic chain is overused or blocked by the absence or excess of intermediary enzymes, imbalance ensues among of the three categories of adrenal hormones.

4.4. The vascular design of the abdomen

The major blood vessels, that traverse the belly, supply blood to the lower body, the spine, and the vital abdominal organs. Like any hydraulic system of fluid flow, the abdominal vascular system is impacted by the belly affairs in its ability to ration its blood flow through the many tributaries and to the different organs. Intestinal compression affects blood vessels in proportion to their length, thickness, exposure, course, and destination. The minute blood **capillaries** suffer most from intestinal disturbances. Those affect structures sensitive to high blood demand such as the **nerves** and **muscles**. Thicker blood vessels such as the **arterioles** are next in getting squeezed by intestinal disturbance. Those affect fleshy structures such as the uterus, bladder, ovaries, testicles, kidneys, liver, stomach, and pancreas.

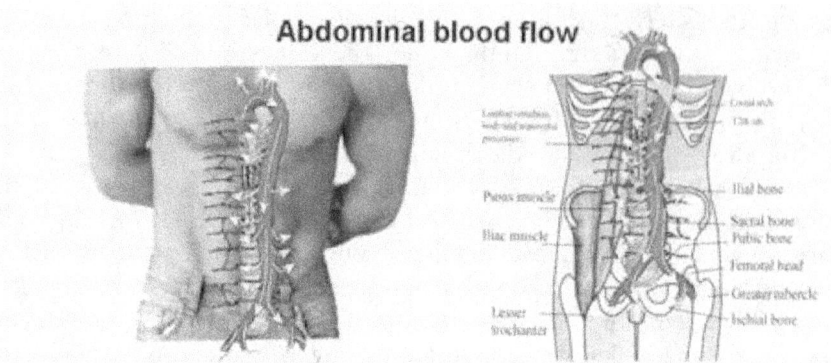

Abdominal blood flow

Figure 12. Blood flow from the aorta to the abdominal organs, lower limbs, spinal structures, and genitalia.

Of course, the veins are at the greatest risk of being obstructed by the increase in intestinal pressure. That is due to the lack of muscular support in the walls of the veins and the low pressure of the blood flow inside veins. **Varicose veins** of the testicles that cause male infertility is related to the congestion of the testicular veins as it traverses the belly, and enters the major venous blood return. Testicles atrophy when their venous blood backs up, which in turns, prevents the flow of nutrients from the arterial blood to the cells of the testicles.

Even though varicose veins of the lower limbs are not directly caused by abdominal distension, they are worsened by it. Further, the deep veins of lower limbs become congested when the belly fails to change its pressure by rhythmic contractions. When abdominal disturbance lingers for decades, the highly pressurized belly cuts the return of the blood from legs and pelvis to the heart, and thus leads to the wasting of the muscle mass of the buttocks, thighs, hamstrings, and calves.

****

As such, a simple condition of poor belly hygiene that ensues due to inadequate mixing of food with the digestive intestinal juices, leads to such system reaction of increased blood pressure. This would require medication which in turns worsens the vicious circuit of intestinal distension and enlargement unless discreet reduction in the quantity of food intake in exercised.
****

CHAPTER 5: THE ABDOMINAL SAC

5.1. The need for pressure change

The softness of the belly is mandated by the laws of mechanics which require **change in volume** in order to produce change in pressure.

Thus, all animals that possess rigid frames, such as vertebrates, must be equipped with an element that could change its volume over time in order to generate change the pressure of its fluids as gases. The pressure change causes **flow**. The flow of fluids and gases enable the aggregated cells of the organism, the human organism, to do the business of sustaining life. That is the delivery of energy between organs and the elimination of waste through the filter.

Therefore, the belly's main role is maintaining, regulating, and **inducing pressure change**. That requires strong and toned muscles. The change in the pressure within the belly changes blood flow, air exchange, and fluid clearance. Like any fluid pump, the belly requires bone support in order to initiate pressure against. The belly forms complete sac attached to the spine, either directly through the abdominal muscle proper, or through the **chest ribs** and **pelvic bones**.

5.2. Spinal attachment of the abdominal muscles

The **direct abdominal attachments** to the spine, on both sides of the spinal vertebras, tend to bend the spine in the direction of fetal posture (head approaches knees). These group of abdominal muscles are perfect squeezer of the interabdominal contents such as the **intestines, kidneys, blood vessels, liver, spleen, uterus, and bladder**. Each of those abdominal organs is affected by the belly hygiene despite the fact that constipation is the commonest complaint of poor abdominal health. The most damaging effect of poor abdominal health, however, is the compression of the kidneys that leads to high blood pressure. That is a major killer in modern societies.

Before you get puzzled, recall that **high pressure** differs from **change in pressure**. Healthy belly maintains rhythmic abdominal pressure tailored to healthy life style. By that I mean:

1. Regular intestinal evacuation,
2. Comfortable breathing on exertion,
3. Adequate fluid clearance, and
4. Robust low back strength.

Poor belly hygiene undermines the rhythmic pressure change and leads to increased abdominal pressure due to **accumulated intestinal contents**, infrequent and incomplete defecation. The congested and overburdened intestines lose muscle tone, dilate, enlarge and aggravate the evacuation process. A vicious circuit ensues by weakening the abdominal muscles due to the protracted over distended intestine. The weakened abdomen caused kidney stress by both direct compression on the kidneys and blockage of urine flow. The stressed kidneys trigger high blood pressure due to inadequate clearance of fluids. The elevated high blood pressure aggravated the general health of the individual. The vicious circuit continues.

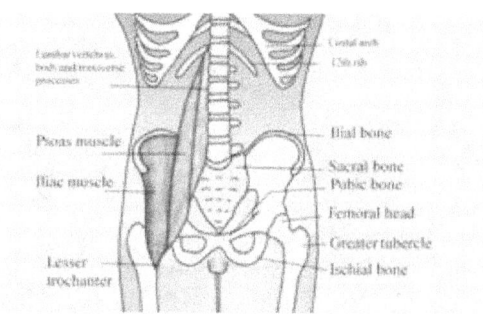

Figure 13. The abdominal muscles comprise an enclosed sac that contains three different systems. Those are: (1) The digestive system which utilized chemical reactions and bacterial growth in order to breakdown food, absorb its nutrients, and store its remains, (2) The urinary system is a sterile filtration system that manages the chemical balance of the blood molecules; and (3) The vascular system transports great amounts of blood through the belly into the genitalia and the lower limbs. The abdominal sacs stretches between the pelvic bones (ilium and pubis), the chest cage (ribs and sternum), and the spine. The belly proper has three walls: muscular abdominal sac, diaphragm comprising the roof of the belly, and pelvic abdominal wall comprising the floor of the belly.

5.3. Other attachments of the abdominal muscles to bones
The **indirect abdominal attachments** to the spine, through the chest ribs and pelvic bones, serve specific goals. The abdominal attachments to the chest cage create the **accessory respiratory** abdominal muscles that affect both the abdominal functions and the cardiopulmonary functions. The abdomen exerts its pressure change on the chest cage through the pull and release of the ribs and the contraction and relaxation of the diaphragm. The accessory pressure change, lent to the chest by the belly, facilitates mobility that requires great change in pressure of body fluids. In the floor of the belly, the abdominal attachments to the pelvic bones serve other greater functions. The **abdominal floor** assists in defecation, ejaculation, delivery of fetus, and regulation of blood flow to the legs, genitalia, and low back. Externally, the abdominal attachments to the pelvic bones enable us to turn and twist of our hips around the spine.

While the soft belly gains mechanical support from its attachments to the bones of the spine, pelvis, and chest, it strengthens those bones by both stimulating mechanical pull and **mobilizing actions** and by pumping fluids that **deliver nutrients** needed for the strengthening process. Poor belly hygiene undermines both processes of mobilization and flow of nutrients and thus leads to weak abdominal muscles and stiff and thin bones.

The main goal of belly hygiene is the maintenance of flow between fluid energy and solid energy, with the **blood flow** comprising fluid energy, and **bone densification** comprising solid energy. It should always be remembered that strong bones are signs of healthy pumping of minerals into the structured matrix of bones by healthy biological system. In other words, strong bones are rich solid fuel which we rely upon as a continuing reserve of mineral supply. When bones are weakened, muscles lax, the skeletal frame crumbles, respiration is hampered, circulation troubled, and general health declines. Thus, healthy lifestyle is founded on promoting bone strength and maintaining resilient blood flow that could assist individual in enjoying active living.

Therefore, the belly's main role is maintaining, regulating, and inducing pressure change. That requires strong and toned muscles. The change in the pressure within the belly changes blood flow, air exchange, and fluid clearance.

5.4. ANATOMY OF ABDOMINAL MUSCLES

The abdominal sac is enveloped by the following four muscles:

o The Rectus abdominis in the front, *Figures 14 and 18*, originates from pubic crest of the pelvis and inserts into lower front ribs and the xiphoid process of the chest. It runs in the front of the belly and flexes the spinal axis by pulling the chest cage to the pelvis. It carries the famous 6-pack appearance in slim people.

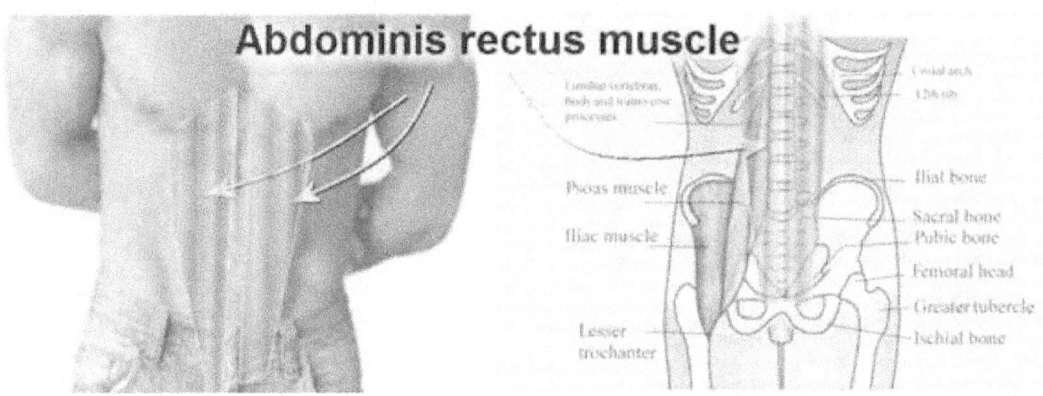

Figure 14. The abdominal rectus

o **Obliquus externus abdominis** on the surface of the flanks, *Figures 15 and 16,* originates from the lower front and back ribs, lumbar fascia, ilium, and Inguinal ligament and inserts into the fascia of Rectus abdominis, ilium, inguinal ligaments, pubic crest, conjoint, and ribs, and linea alba.

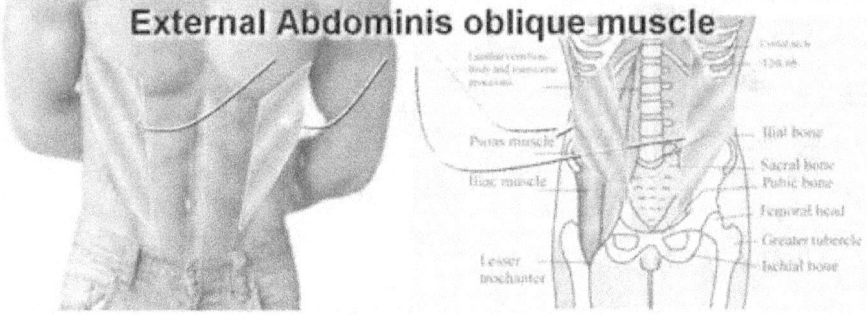

Figure 15. The external abdominal oblique.

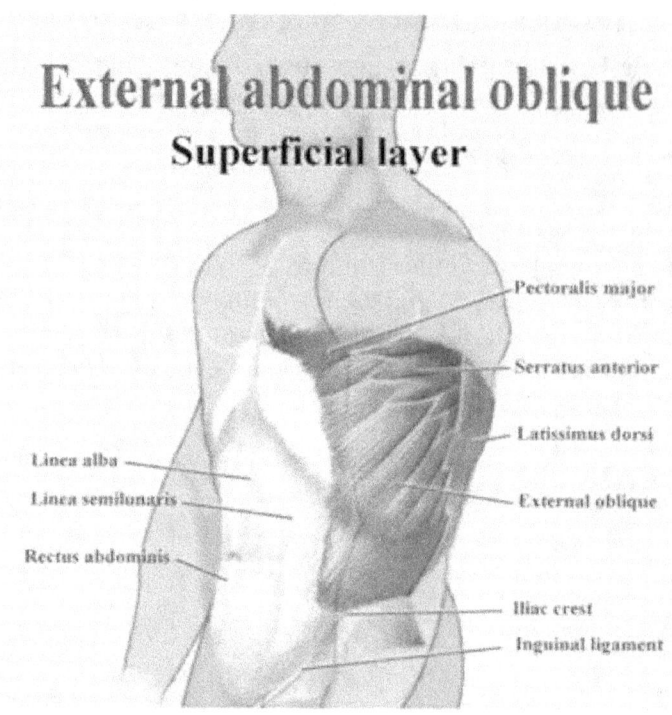

Figure *16.* Obliquus externus abdominis muscle.

o **Obliquus internus abdominis** under the externus and on the flanks too, *Figures 17 and 18*, also originates from the lower front and back ribs, lumbar fascia, ilium, and Inguinal ligament and inserts into the fascia of Rectus abdominis, ilium, inguinal ligaments, pubic crest, conjoint, and ribs, and linea alba. It runs perpendicular to the obliquus externus across the sides of the belly. Both oblique muscles pull the chest cage to the side towards the pelvis by flexion and rotation by way of their perpendicular direction.

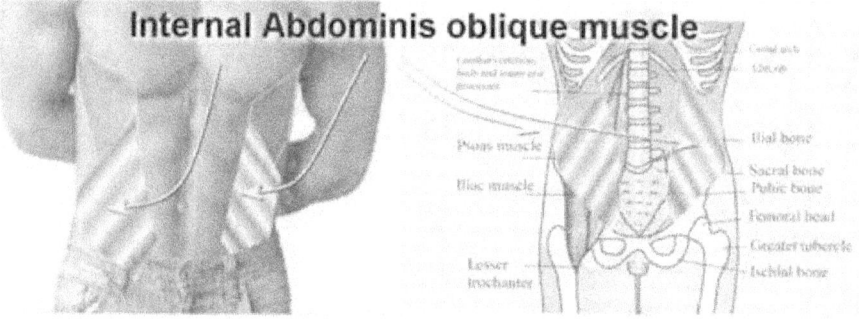

Figure 17. The internal abdominal oblique.

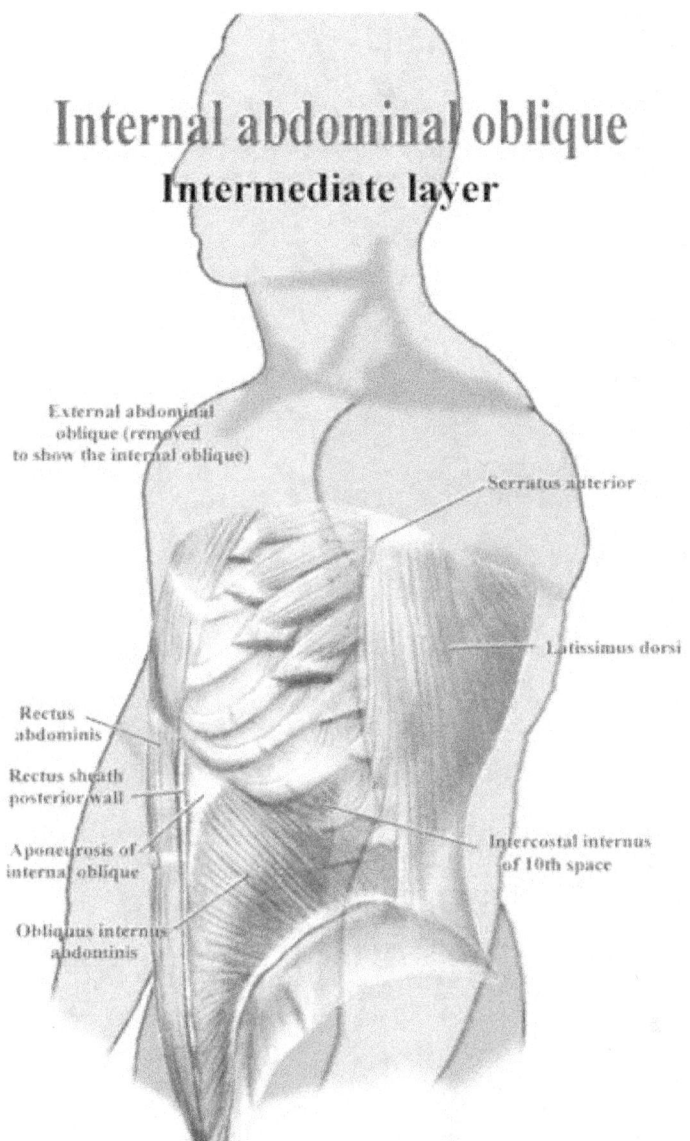

Internal abdominal oblique
Intermediate layer

External abdominal oblique (removed to show the internal oblique)

Serratus anterior

Latissimus dorsi

Rectus abdominis

Rectus sheath posterior wall

Aponeurosis of internal oblique

Intercostal internus of 10th space

Obliquus internus abdominis

Figure 18. Obliquus internus abdominis muscle.

o **Transversus abdominis** deep to the internus and on the flanks, *Figures 19 and 20*. This muscle pulls the rectus sheath of the Rectus abdominis inwards, towards the vertebral column thus moves the naval inwards. It thus increases the intraabdominal pressure and could move the guts upwards.

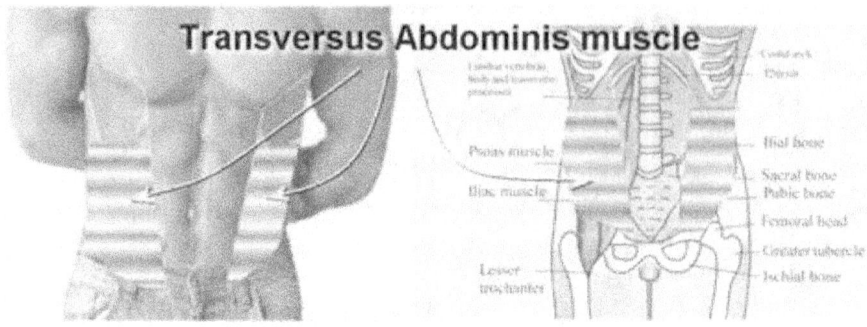

Figure 19. The abdominal transversus.

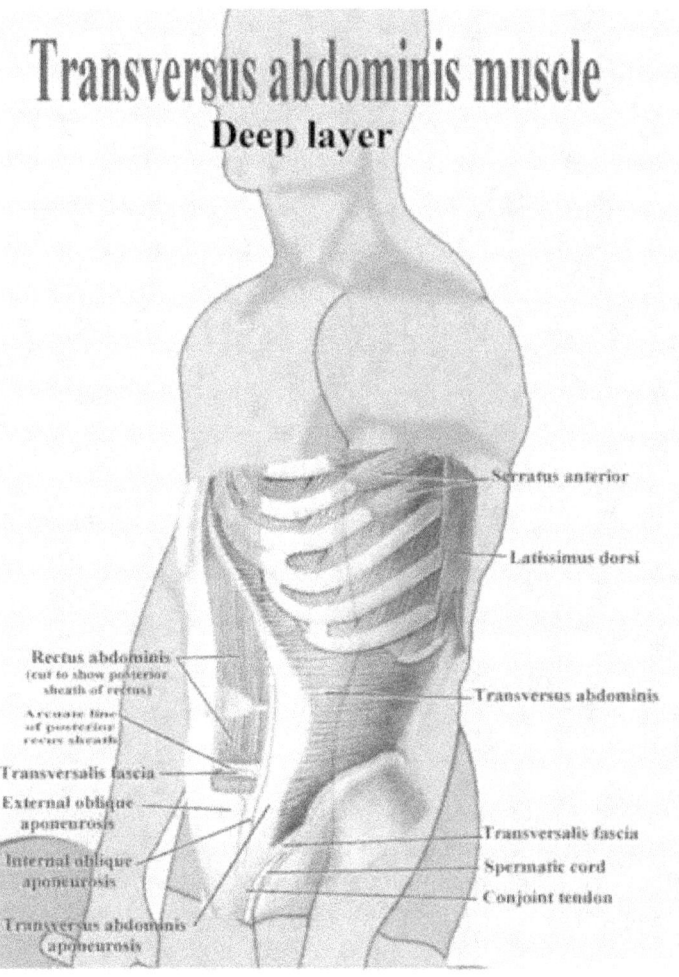

Figure 20. Transversus abdominis muscle.

5.5. Abdominal filling defects

The outer shape of the belly is determined by its contents. Gases must be contained within the intestinal cavity in normal health. If the intestine leaks gases to the outer space, the peritoneal cavity, emergent surgical procedure would be required in order to contain the septic spill from the intestine to the neighboring structures. Intestinal distension causes **round abdominal contours**. The gas filled belly is easily confirmed by its tympanic percussion resembling a drum.

Fat deposits confer irregular contours on the belly due to the semi-solid nature of fat that prevents uniform distribution over the belly. Fat deposits wherever the cellular spaces lack access to efficient burning mechanisms by mitochondria. The **population of mitochondria** depends greatly on the physical activity of the organs and structures. The buttocks, belly, and thighs are the least active structures in individuals living sedentary lifestyle.

In growing children, the belly bulges in front of the body due to the premature development of the bones of the pelvis of the child. As the child grow, the **pelvic cavity** becomes roomier, and belly retreats as its contents descend into the pelvic cavity.

The **pregnant belly** is filled with the soft tissues of the fetus, placenta, and the enlarged uterus that supports perception. Fat in a pregnant belly must have been deposited prior to perception since pregnancy limits the ability of the individual to consume beyond her needs.

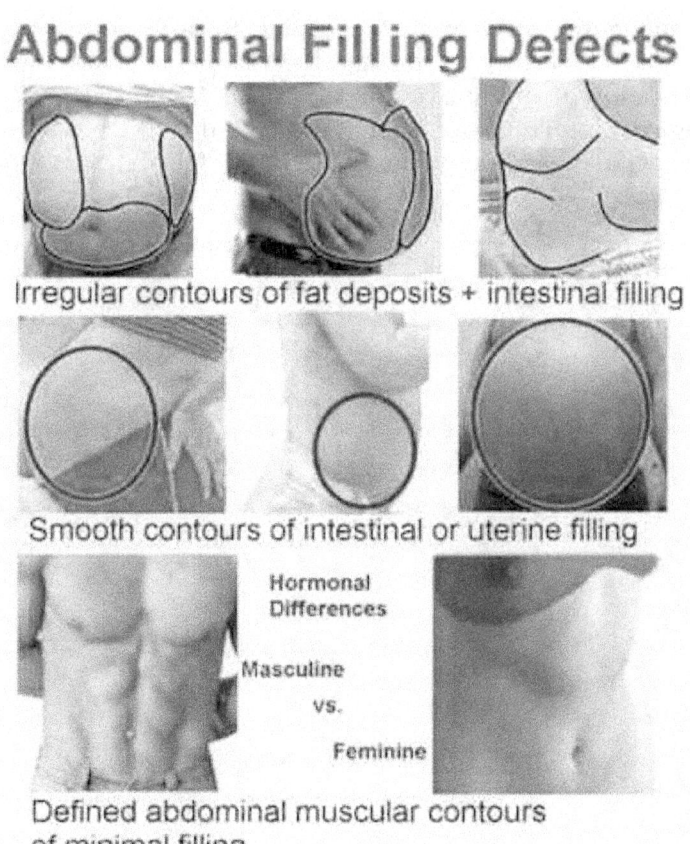

Abdominal Filling Defects

Irregular contours of fat deposits + intestinal filling

Smooth contours of intestinal or uterine filling

Hormonal
Differences

Masculine

vs.

Feminine

Defined abdominal muscular contours
of minimal filling

Figure 21. The abdominal contours reflect the nature of the filling defects of the belly. The nature of the abdominal filling defects determines the outcome of the general health of the individual. The fat belly is notorious in creating creases that are hard to cleanse. Further, as fat deposition interferes with breathing and circulation, the individual loses his/her ability to attend to personal hygiene and care of skin. **Bacteria, fungi, and viruses** invade the poorly circulated and poorly cleansed fat creases through various routes. From the anus, mouth, and nose, opportunistic organisms spread by local **secretions**, carried in the **nails** of person during scratching, and through **contaminated clothes**. The inability of the individual, to bath daily and efficiently and change dirty clothes, leads to opportunistic skin infections that vary from minor irritation to life-threatening **cellulitis** and **septicemia** (spread of infection to the blood).

The most overlooked aspect of health care of obese people is the scarcity of resources in face of the excessive weight gain of patients. A patient weighing three hundred pounds would require from **three to four nurses** to care for a single accident of fecal spill in a hospital sitting. Unfortunately, there are abundance of patients weighing over five hundred pounds with whom hospitals are bombarded on daily basis. Neither the human resources, nor the technical equipments of any hospital suffice to attend to the health crises of such overweight population.

****

From the anus, mouth, and nose, opportunistic organisms spread by local secretions, carried in the nails of person during scratching, and through contaminated clothes. The inability of the individual, to bath daily and efficiently and change dirty clothes, leads to opportunistic skin infections that vary from minor irritation to life-threatening cellulitis and septicemia (spread of infection to the blood).

****

CHAPTER 6: ABDOMINAL EXERCISES

6.1. CRUNCHES

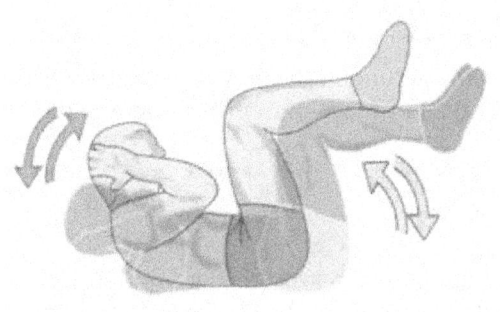

Figure 6.1. Floor abdominal crunches.

Purpose

o Raising the head and chest, while lying on the back, and maintaining buttocks and lower back stuck to the floor or bench, works out the Rectus abdominis and front muscles of the neck.

o Remember that flat belly requires more than strong muscles. It requires less distention of guts and proper transit of contents. These all depend on dietary habits and aerobic activities. The strong abdominal muscles only contract when you want then to do so. Yet, high tone requires habitual low abdominal pressure and more general mobility than just executing few sets of exercises.

Start position

o Lie on your back on a flat bench or on the floor. Both hands are kept behind your head or on the chest, with or without weight, ***Figure 6.1.***

o Both legs are bent at knees and hips or kept straight. Feet can be supported under feet support in specially designed crunch boards, or by letting a partner hold them while raising your upper torso.

o Start exercise by hardening your abdomen, thrusting your chest and supporting your head.

Execution

o Raise your head and chest from horizontal. Your range of motion will not be great since the Rectus abdominis only maintain flat belly. It works mostly by isometric contraction to maintain postural flat abdomen.

o You are bending your upper torso to bring the chest cage towards the lower half of your belly, not the thighs, ***Figure 6.1.*** Thus, you will not get up by using the Iliopsoas.

o The oblique abdominal muscles can either work out by performing sit-ups on the side of the torso or on a chair.

o You can also perform isometric contraction by slightly elevating the head and chest, without apparent raising, and then relaxing them back.

o Reverse the motion after a short or no pause, at the end of raising. As with other exercises, you need to repeat that set from three to ten times or more.

o You can superset crunches with hip flexion by raising both upper torso and legs, while maintaining your lower back abutting the floor.

****

Remember that flat belly requires more than strong muscles. It requires less distention of guts and proper transit of contents. These all depend on dietary habits and aerobic activities.

****

6.2. LEG RAISES

The four abdominal muscles could be worked out together with scapular and hip flexors by performing Leg Raises over a bench. The following figures illustrate the use of a simple flat bench in working out the abdominal muscles to their best.

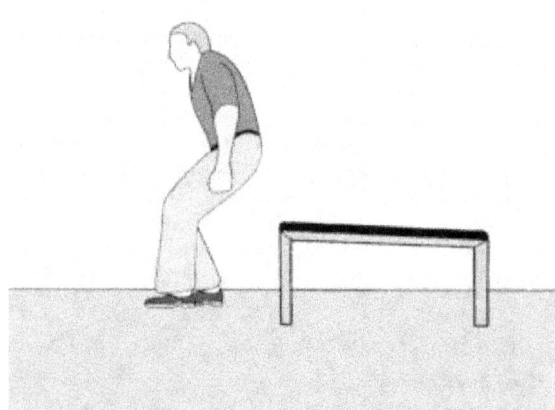

Figure 6.2. Crash back landing: This intensive abdominal exercise requires thorough warm-up and high level of coordination.

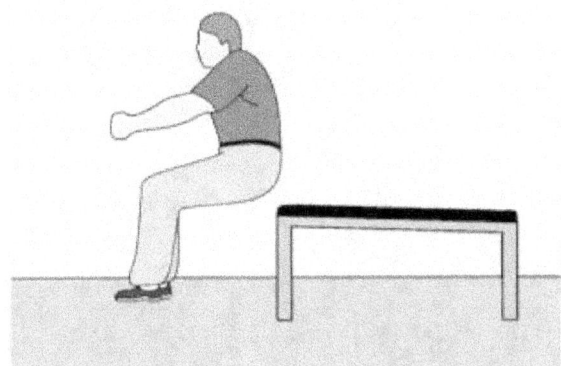

Figure 6.3. Lying down could be softer than just slamming your weight against the bench. But when you are warmed up, you should be able to maintain high coordination.

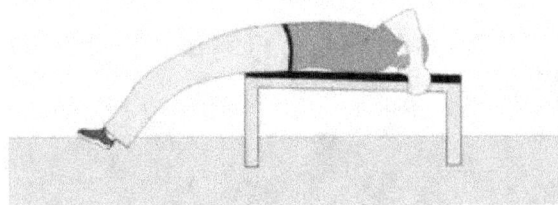

Figure 6.4. Scapular hardening: Here is the toughest part of holding the edges of the bench and tightening your scapular muscles vigorously. These muscles elevate your whole body over the shoulders.

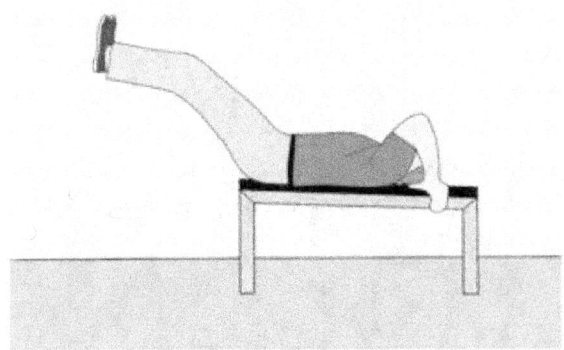

Figure 6.5. Leg raises: If the pelvis does not rise off the bench, then only the Iliopsoas is working out on the legs. Unless you tightening your abdomen isometrically.

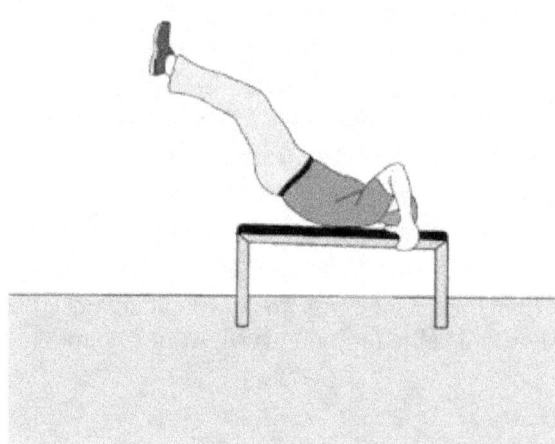

Figure 6.6. Leg and lower torso raises. This is peak spinal flexion. Although the Latissimus dorsi elevate the pelvis on the fixed upper arms, the Rectus abdominis and Iliopsoas share great deal of resistance.

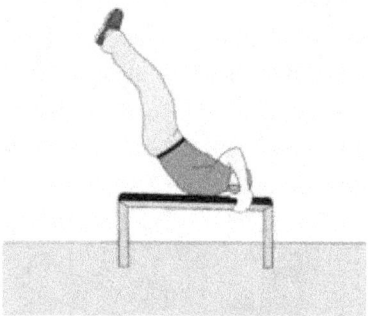

Figure 6.7. Further torso raising is possible to execute with the pectoral and abdominal muscles in the front, instead of the Latissimus dorsi and Triceps in the back.

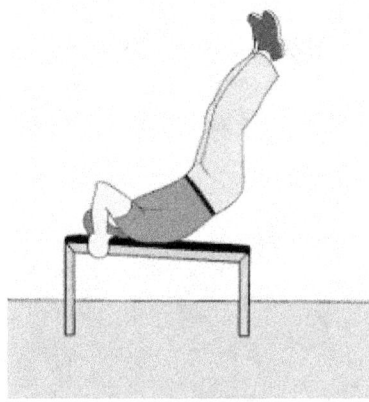

Figure 6.8. Twisting the pelvis on the spine could also be done either by the Obliquus abdominis and pectoral muscles, or by the Latissimus dorsi and Triceps muscles.

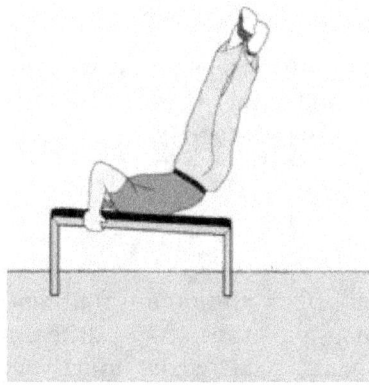

Figure 6.9. Reversing pelvic twisting could be done without lowering the legs. Many repetitions of pelvic twisting could be done in this strenuous posture in order to maintain great proportion of waist to chest ratio.

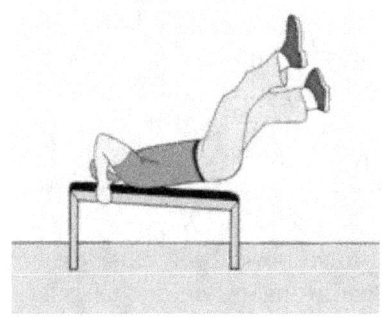

Figure 6.10. Drawing circles in air by the tips of feet, while elevating the lower body on the shoulder plates, rocks the lumbar spines during intense abdominal contraction.

Figure 6.11. High shoulder stand relieves the anterior abdominal muscles as well as the pelvic floor.

****

Another great winner. Here, your entire body is lifted by your Biceps and resting on your shoulder plates. This is the exercise that confirms your axial integrity. Simply, balanced shoulders, spine, and pelvis enable you maintain such integrity.

****

6.3. STRATEGIES OF ABDOMINAL EXERCISE

Most people exercise their abdomen with the fixation of "**six-pack abdominal muscles**" in their minds. For that reason, bodybuilders are often doing numerous sets of different abdominal exercises, with no break in between. Some of them are fanatically doing very long abdominal trainings every day. Generally, however, individuals who exercise a lot, are genetically gifted, and with low body fat percentage, could have such an appearance. Even with a lot of exercises, and genetics on one's side, achieving and keeping very low body fat percentage for lifetime is very suspicious, considering how restrictive, expensive, and impractical for average recreational person that practice could be.

Although people naturally strive for good looking, the main reasons for abdominal training should be quite different than appearance only. Axial training, which is the efficient means for inducing body power, strength, and balance, utilizes lower back muscles heavily, which requires equally toned abdominal muscles in order to support the anterior surface of the spinal vertebrae. Also, the Weightlifting practice shows the need to pull the guts inside by the help of abdominal muscles in order to level the **pressure inside abdominal cavity**. Abdominal muscles are partly responsible for the proper upright body posture (see section 6.3.1). Apparently, there is a balancing interaction between abdominal and lower back muscles in strength exercises and in common life.

Abdominal exercises should be carried out preferably with fresh lower back muscles. For that reason abdominal exercises should precede heavy Deadlifting or similar exercises.

For most people, the best possible time for doing abdominal exercises is probably the early morning just after getting up from bed. There are good reasons for such a training however. These exercises mostly do not impose the high demand on the spine. They are good for controlled and gradual warming up of the body. Yet, there still would be an additional healthy benefit about such early exercises: the **intestine effect**. The regular and natural intestine stimulation and cleansing is best done just with morning exercising that engages abdominal muscles. There are several positive physiological and psychological effects of such early morning training:

• Regular intestine cleansing is very beneficial from the medical point of view;

• The stomach and intestine are reset for daily trainings and routines;

• The individual's mind gets prepared for the forthcoming efforts both in training and in other daily activities.

****

For most people, the best possible time for doing abdominal exercises is probably the early morning just after getting up from bed.

****

6.3.1. THE PROPER UPRIGHT POSTURE AND ABDOMINAL MUSCLES

The strong abdominal muscles are important for balancing the **shearing forces** acting upon the lumbar region in the upright body position. The proper upright posture is sometimes explained to the children at schools. Elders however usually do not pay much attention to that advice, at least until they sense the sharp pain in their lower back. Sometimes the pain is felt in the lumbar region during long strolls, returning home after a very hard training, unexpected step down from a sidewalk or just when the waist body muscles get tired enough and the body is in the prolonged upright position. Thus, it is good to know the proper posture described as follows:

- shoulders pulled back and down;
- bellybutton pulled towards the spine and up;
- lower back straightened up;
- ears, shoulders, hips and knees are on in plane.

Proper posture alleviates breathing because of **thrusting of chest cage**. When assuming the proper upright posture, one is inducing the better breathing and better counterbalancing of the shearing forces upon the spine. The individual would be fond of finding that visual body appearance is also improved when keeping the correct upright posture. Finally, the straight body position induces better personal feelings than bowed and hunched position.

The proper posture helps in preventing pain in the lower back if it is caused by the tired waist or shoulder muscles. By keeping the correct upright posture, individuals are actually strengthening their his muscles in the process. If keeping the proper upright posture gets too hard, it is usually a sign of mental and/or physical conditions that require rest, or requires the strengthening of the back, shoulders or abdominal muscles.

****

Proper posture alleviates breathing because of thrusting of chest cage. When assuming the proper upright posture one is inducing the better breathing and better counterbalancing of the shearing forces upon the spine. An individual would be fond of finding that visual body appearance is also improved when keeping the correct upright posture. Finally, the straight body position induces better personal feelings than bowed and hunched position.

****

6.3.2. VARIETIES OF ABDOMINAL EXERCISES

The abdominal exercises proposed in this section are divided into three sections: **basic** abdominal exercises, abdominal exercises **with horizontal bar**, and abdominal exercises **on bench**.

6.3.2.1. Basic abdominal exercises

The following basic set of abdominal exercises is carefully chosen so individuals can perform the exercises with no additional machines or any other aids. All that is needed is just one appropriate mat to lie upon. These basic exercises can be carried out on in the sequence just as presented.

The basic abdominal exercises are very well suitable for regular morning training because they do not need any gym accessories. Consequently, these exercises are good for every lifestyle.

In the described exercises, the term "**upper torso**" refers to the chest and the shoulders. When the upper torso is raised, I assume that you will also raise the arms and head with it. Also, the term "**hip**" is used in many instances here to refer to the pelvic region, not merely the hip joint. Most exercises in this group commence from lying **supine** (on your back). When lying on your back, you should generally strive for the straight lower back, bellybutton pulled towards the spine, and avoid any protracted **lordotic** (curved inside) position of lower back.

****

Though the following aggregations of exercises appear extensive, redundant, or overwhelming, they provide a wide scope of options to choose from. The more you read about the anatomy of the abdominal muscles and spines, the easier you remember the implementation of all combinations of relevant exercises.

****

1

1. Starting position: while lying on your back, cross your arms over chest or keep them stretched above your head, with hand in hand. Knees are bent perpendicularly and shins are parallel to the ground.

Action: bend your torso towards your hips, while elevating your hips off the ground; with both knees bent, shins parallel to the ground, and thighs perpendicular to it. Optionally, while in peak contraction put your hands over your knees.

2

2. Imagine climbing a rope while lying on your back, with your thighs perpendicular to the ground. Raise one hand on the imaginary rope along with your hip on the same side of that hand. Lower the raised hand and hip. Repeat the same on the opposite side. Variants: change the angle of your knees and feet in one set or between consecutive sets. The raising of the shoulder and hip on the same side contracts the Rectus abdominis on that side alone.

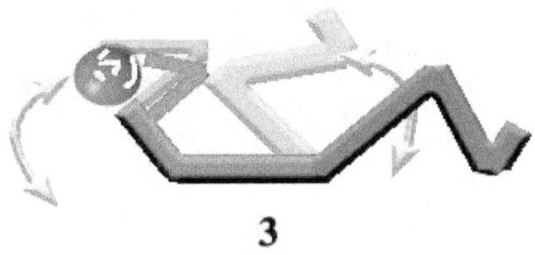

3

3. Starting position: while lying on your back, keep your feet in the air just above the ground. Arms are behind the head; lower back is flat; pull bellybutton towards the spine; both knees bent, as shown in the picture.
Action: bend your torso towards your hips, while elevating one knee towards your head, keep contraction for a while and then return to the starting position. Repeat on the other knee.

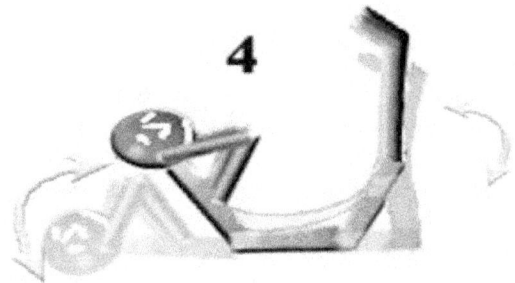

4. Elevate both upper torso and hips while keeping your legs high in the air. This exercise must be done steady and not too fast. If the exercise were done fast, the lower back would receive jerks while in kyphotic position (curved outwards). The raised legs stretch the hamstrings and the spinal erectors and add to your mind the ability to exercise your abdominal muscles when the back extensors are stretched.

5. Similar to the exercise 4. Steady frequency and moderate speed is crucial. The extended arms shift the center of mass and contract the Deltoids. Thus, this exercise adds more complexity to the combination of muscular actions by combining arms and legs extension with spinal flexion. Remember that all these combination of muscular actions enrich your motor control over your body.

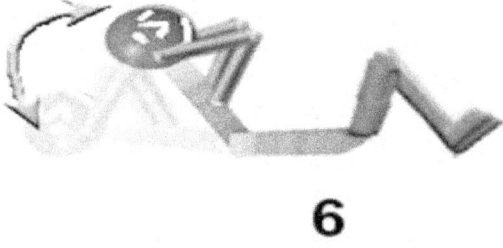

6. Keep your lower back all the time straight and stuck to the ground. Keep your heels on the ground and toes in the air. The bent hips relax the Iliopsoas muscles.
Action: raise your head, shoulders, and chest upwards. This mainly contracts the Rectus abdominis muscle.

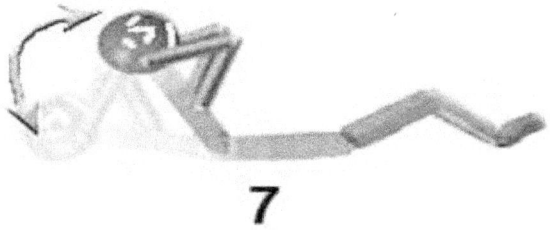

7

7. Legs are almost straight with knees slightly bent. This stretches the Iliopsoas muscle and thus pulls the lower vertebrae upwards.
Action: raise your head, shoulders, and chest upwards, like the previous exercise.

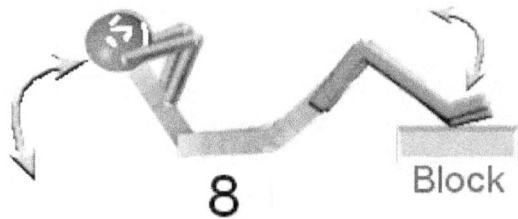

8   Block

8. Raise both torso and hips while using a chair, bad or sofa to elevate your feet. This alters the starting point of the Rectus abdominis since the raised legs rounds the lower spine and shorten the distance between the ribs and the pubis symphysis.

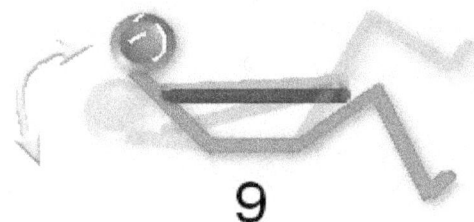

9

9. Keep your feet in the air close to the ground. This tenses the Iliopsoas and develops its endurance while contracting Rectus abdominis. It tightens the lower spine since raising the legs is an active contraction of the Iliopsoas.
Action: raise your head, shoulders, and chest upwards, like the previous exercise.

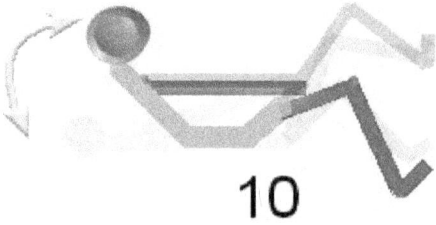

10. Keep your heels on the ground and feet toes in the air while elevating both the hip and torso.
Elevating the hips contracts the Iliopsoas. Elevating the torso contracts the Rectus abdominis.

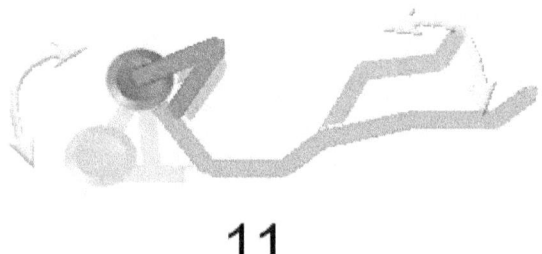

11. Do one set slowly and the other set fast. Do not forget to return your scapulas to the ground. The hands on the head shift the center of mass towards the head and makes raising the torso more difficult.
As the previous exercise, this one contracts both Iliopsoas and Rectus abdominis.

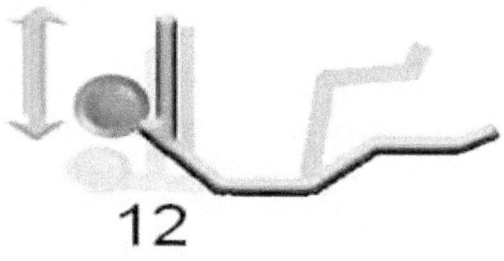

12. Similar as exercise 2 but with legs spread apart, with similar variations.
The spread femurs pull the insertions of the Iliopsoas muscles laterally and thus eliminate any hip elevation unless the Rectus abdominis pull the pubis symphysis upwards.

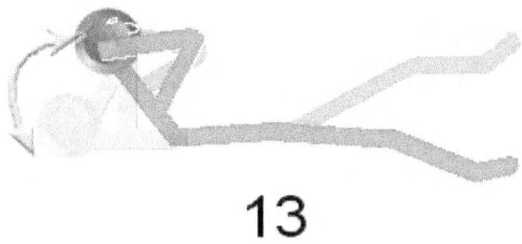

13

13. Both heels are on the ground, with legs just slightly bent in knees. The spread legs and extended hips pull the lumbar spine anteriorly and anchor the hips on the ground. This is pure Rectus abdominis exercise since the head, shoulders, and chest are elevated with the pulling of the lower ribs and sternum towards the pubis symphysis.

14

14. Raise your upper torso and hands towards the ceiling with legs spread apart and high in the air. This balances spinal flexion with Rectus abdominis resisting the weight of the head, shoulders, and chest against spinal extension with the Iliopsoas resisting the weight of the abducted legs. Notice that though the Iliopsoas flexes the hip joint it extends the lumbar spine by pulling them forwards.

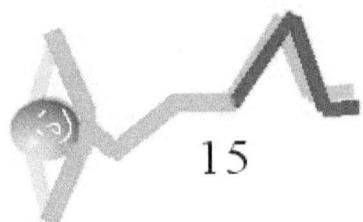

15

15. Bend both knees and position the feet on the ground half way to pelvis. Twist your upper torso at your waist to one side. Raise your twisted upper torso from the floor so your higher elbow points towards your thigh. Keep that contraction for a moment while feeling your oblique muscles doing the work and pulling your hip towards your shoulder. Then do the same on the opposite side.

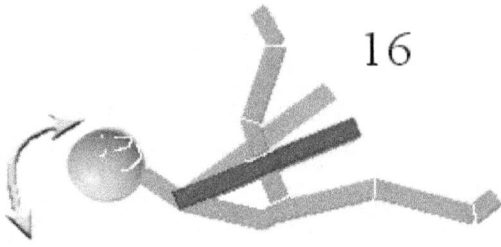

16. Starting position: when lying on your back, keep the arms close to the sides of the body with both legs slightly bent in their knees and heels barely touching the ground.
Action: swiftly raise both your upper torso and one of your legs high in the air, aim your arms around the raised leg; the lowered leg is also slightly elevated close to the ground. Then swiftly lower both your torso and raised leg to the ground while pressing the heel of your lowered leg against the ground. You are actually using your grounded leg as cushion, relieving your lower back of the pressure and helping you keep the balance at the same time.
Do the same on the opposite side in the next repetition or in the next set.

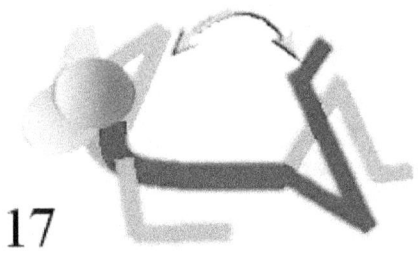

17. Keep one hand on the stomach while the elbow of the other arm is flexed. The upper torso raises and rotates such that the high elbow follows a broad movement from touching the ground on its side, to touching the opposite knee. You can change the frequency of the exercise as you wish but do not shorten the movement range. This contracts the Obliquus externus abdominis on the side of the mobile elbow and the Obliquus internus abdominis on the opposite side.

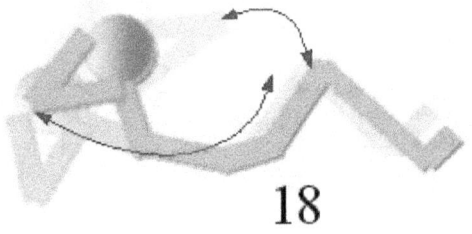

18. While raising head, shoulders, and chest from the ground, twist your torso to the side simultaneously with pulling pelvis towards the same side. Repeat on the other side.

This is a combined workout for the Rectus abdominis and Obliquus abdominis muscles. Also, maintaining the waist size smaller by contracting the Transversus abdominis adds another benefit.

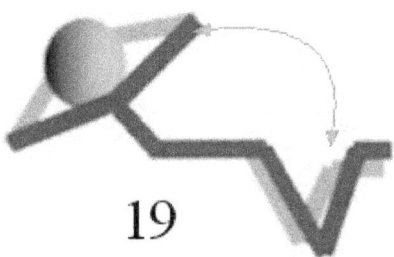

19. At the start, keep both legs together, bent at the knees and hips, and with both knees lowered to the ground on one side. This posture positions one side of the pelvis on the ground and brings the insertion of the Obliquus abdominis muscle into the line of pulling of the upper torso from the floor.

Place your hands behind your head and the shoulders parallel to the ground.

Raise your upper torso upwards with the shoulders still parallel to the ground. Variant: you may try to raise your pelvis a bit during raising the upper torso.

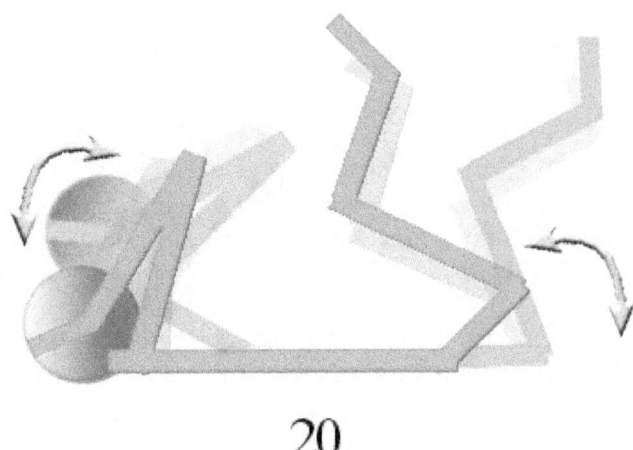

20. This is the most common exercise for the Rectus abdominis that utilizes the weight of the lower body to strengthen the abdominal muscles. If you tighten the abdominal pouch

all around, you also strengthen the Transversus abdominis muscles, which pull the rectus sheath posteriorly.

Elevate your lower body from the ground, including the buttocks. If the buttocks do not rise above the ground then you are strengthening only the Iliopsoas. In order to stimulate muscles you can combine this exercise with the next one.

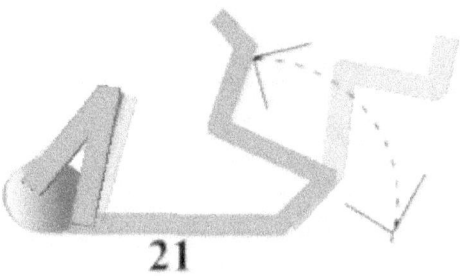

21. Starting position: legs are bent in knees and feet in the air, just close to the ground. While elevating the legs as shown in the figure, maintain tight abdominal muscles all around (front and sides) by suspending your upper torso and head barely over the ground. This is almost a pure isometric tensioning of the abdominal muscles since the rising leg merely strengthens the Iliopsoas and rocks the symphysis pubis up and down.

For better muscle stimulation, combine this exercise with the previous one.

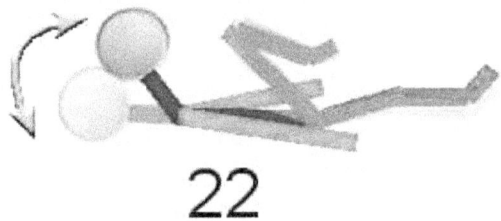

22. In starting position, the straight leg is about one foot above the ground, scapulas are touching the ground and hands are under buttocks. Imagine that you are riding a bike while lying on your back; yet raise the bending knee towards your chest while raising the upper torso. Control the duration of contraction to your need. Then return to the starting position with scapulas touching the ground and both legs straight above the ground. Repeat on the other side. Do not shorten the range of motion during the exercise.

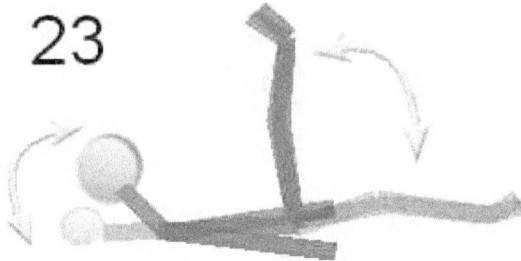

23. From lying supine, you will raise both straight legs and the upper torso off the ground. Do not rush in your way down. Keep the maximum contraction for a moment. When returning to starting position keep the lower back fixed to the ground.

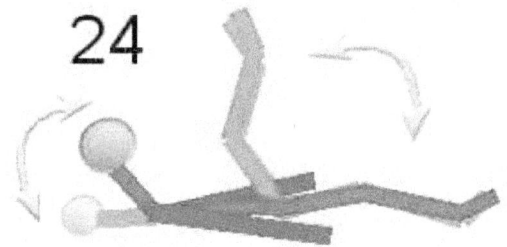

24. As in the exercise 16 but with hands placed under the buttocks. The asymmetric leg motion with upper torso rising imposes impulsive contraction on one bank of the Rectus abdominis muscle, in addition to pulling the lumbar region downwards.

Recall that the Psoas major muscle that raises the leg on the hip joint takes its origin from the front of the lumbar vertebrae and their transverse processes. The contraction of the Psoas major muscle ensures that the lumbar discs are pulled forwards during increased intrabdominal pressure.

25. As in the previous exercise but with hands placed behind the head. This is an added level of difficulty over the previous exercise since you will carry the weight of your arms on your upper torso. Additionally, raising your elbows this way moves your scapulas away from the upper spine and rounds your spinal curvature more than the previous exercise.

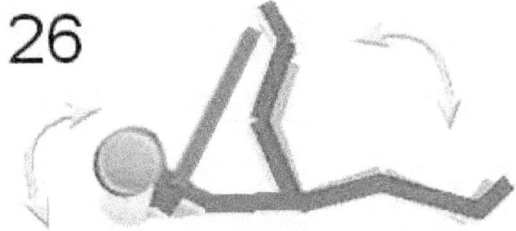

26. As in the previous exercise but with hands extending towards your feet while raising the upper torso. The arm motion from head to toes rocks the upper torso during abdominal contraction. If you have a mean to measure your intraabdominal pressure or your venous blood pressure in the major neck vessels, you will find that that apparently minor arms motion causes waves of pulsation in internal pressure.

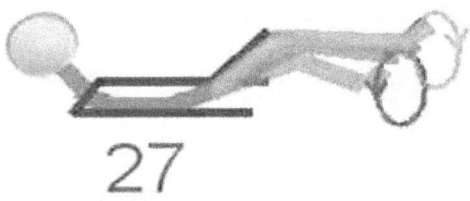

27. With both feet and upper torso raised off the ground, move your feet to make 8s, circles, crisscross, vertical up-down, horizontal side –to-side, or whatever you like. Remember that your lower back should be glued to the ground and bellybutton pulled towards the spine. Do not let the flanks of your abdomen to bulge by tensing it all around. This challenges the Transversus abdominis muscle, which maintains intraabdominal pressure adequate for physical activities. Defects in this muscle lead to hernia.

28. Lie on your back as in the figure, with feet touching at the heels and knees close to the ground as much your groin permits. Feet are all the time on the ground. While tightening your abdomen all around, raise your upper torso only. This challenges the Rectus abdominis. Yet the initial and the sustained tightening of the abdomen challenges the Transversus abdominis as well,

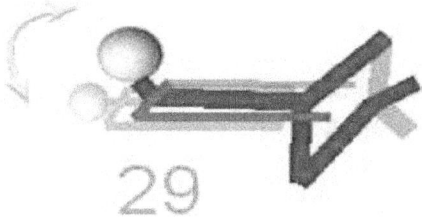

29. Similar to the exercise 28 but feet are in the air just above the ground. When you raise your upper torso, point your arms towards your feet. When lowering your torso just relax your hands above the stomach. With your arms extended in front of you, your upper spine will tend to round unless you thrust your chest forwards.

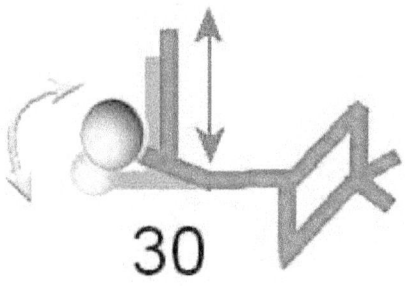

30. Similar to the exercise 29 but here you imagine climbing the rope. Instead of the entire upper torso, you only raise half of it together with your head. Only one shoulder and your head would advance upwards ahead of the other shoulder. This stimulates the abdominal muscles at their origin from the lower borders of the ribs.

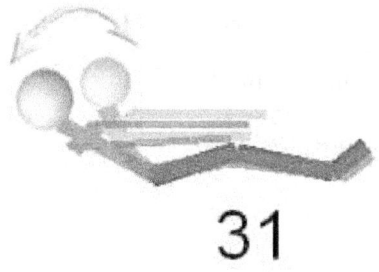

31. In this exercise, you keep the same position shown in the figure for some time, with straight lower back and heels touching the ground. If you want to move your torso more towards the ground, just advance your heels more away from your pelvis. When you want to rest just sit in normal sitting position, with straight lower back. This exercise imposes pure isometric contraction of the anterior abdominal muscles.

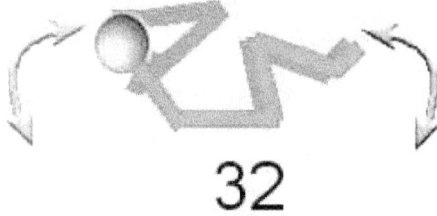

32

32. From lying supine with both legs bent at hips and knees and hands placed behind your head, raise your upper torso and your bent legs simultaneously towards each other. If your waist size permits, you could also elevate your pelvis a bit. This is the famous fetal posture, which squeezes the contents of the guts between the diaphragm and the pelvic floor (the muscle hammock that support the guts against the bony boundaries of the pelvis).

33

33. Starting position: lie on your back with arms stuck close to the sides of the body and with one leg bent in the knee.

Action: raise both your upper torso and the stretched leg high in the air towards each other. At the peak contraction, elevate the hip of the raised leg, while raising the foot of the other leg in the air just above the ground. When lowering your raised leg and torso, put the other foot on the ground. Repeat the same on the other side. This way the grounded leg is used to dampen and ease the pressure on the lower back. The lower leg pulls as a lever on the lumbar spine, via the Psoas major muscle, and thus prevents excessive rounding of the spine at this burdensome posture.

34

34. Starting position: hands behind head, back on the ground, legs spread wide in the air.

Ending position: both upper torso and pelvis raised, knees are touching each other and pulled towards the head. Do this exercise fast but keep the maximum contraction for a few moments.

Remember not to let the flanks of your abdomen go soft, by tightening your waist all around.

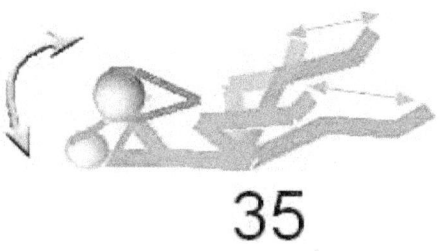

35

35. Starting position is the same as in the exercise 34, but in the ending position, the knees are wide spread and your elbows are touching your knees. When you raise your upper torso, also raise your pelvis a bit.

With abducted thighs, you reduce the action of the Iliopsoas muscle and force the Rectus abdominis alone to pull the pelvis to the chest.

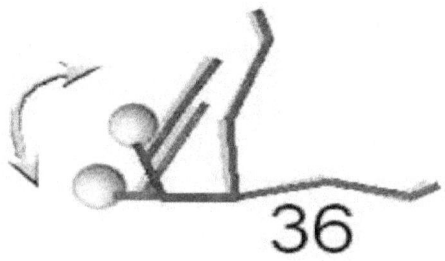

36

36. Keep the lower back stuck to the ground all the time. When arms reach the ground above your head, do not keep them for long. The exercise is done at moderate pace. When raising your arms also raise your pelvis a bit.

The purpose of the wide arms movement is to generate both momentum and compound motion that maintain your readiness to perform unexpected movements.

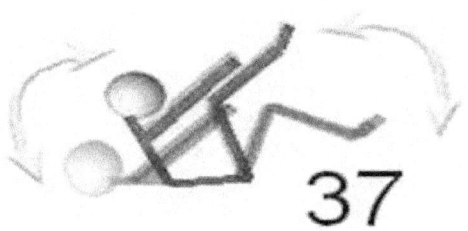

37

37. This exercise is done fast. When raising the upper torso, the hips are raised too. Keep the maximum contraction for a few moments. When lowering the upper torso, the arms touch the ground above your head just for a moment, but the legs are partially lowered to the ground (see the drawing).

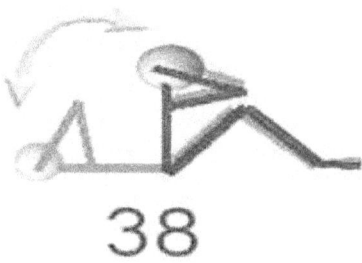

38

38. Raise your upper torso, while legs are in constant bent position at the hips, knees, and ankles. Though you mainly get up by the isotonic contraction of the Iliopsoas muscles, your Rectus abdominis must maintain isotonic contraction in order to keep the upper torso in line with the lower spine.
If you maintain tight abdominal flanks, you challenge the Transversus abdominis as well.

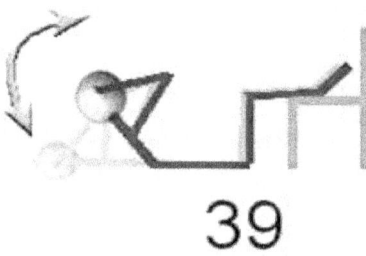

39

39. Anchor your feet under some fixed object so the thighs are kept perpendicular to the ground. Raise your upper torso with your hands crossed over your chest. Preferably, keep some additional weights in your hands to make the exercise more effective. The upright thighs will impose pure vertical pull on the lumbar spine when you attempt to raise your upper torso.

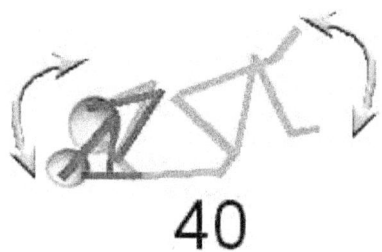

40

40. Place one heel over the other knee and raise the knees and upper torso towards each other. In the meantime, raise your hips a bit. Keep the maximum contraction for a few moments. The locked over leg gives the unusual feeling of the hamstrings trying to push the other knee, while spinal flexion is underway.

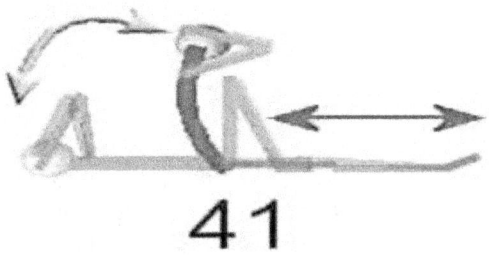

41. Lie down on your back. Sit up with hands behind your neck and legs pulled in flexion. Try to control the tempo of returning to lying down so that your upper torso is lowered slowly and the legs are extended at equal pace.

42. Lie down with hands behind your neck. Raise your upper torso to half-sitting position and legs in half-bent position. When returning to lying down, do not let torso just drop down. Always keep the lower back stuck to the ground to ensure your active spinal erection.

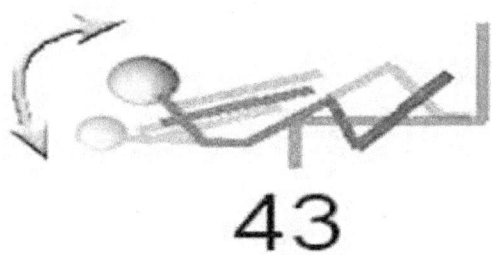

43. Starting position: lie down and place your legs on a chair, bad, or sofa so the thighs are perpendicular to the ground. Spread your knees wide apart and join the feet together. Action: raise the upper torso and the hips at the same time and reach the ankles with your hands.

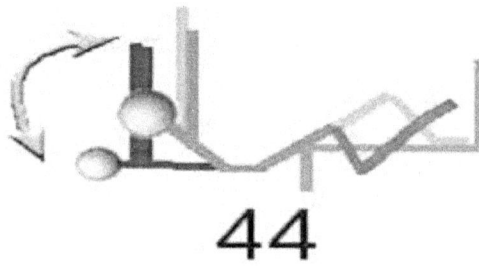

44

44. Similar to the exercise 43 but imagine climbing the rope. The climbing hand pulls one shoulder upwards ahead of the other. Thus, the torso twists and the chest cage rocks from side to side. The Obliquus abdominis muscles do the twisting. They are helped by the Latissimus dorsi and the spinal erectors in elevating one shoulder at a time.

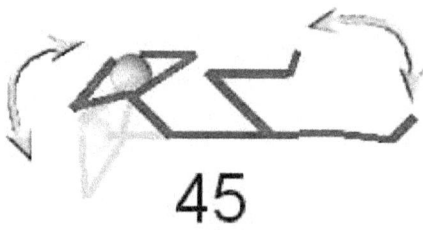

45

Exercise 45: Version 1: touch one knee with the opposite elbow. Lie down. Repeat by touching the other knee with the opposite elbow. This makes for one repetition.
Version 2: touch one knee with the elbow on the same side. Lie down. Repeat by touching the other knee with the other elbow.
Version 3: touch one knee with the opposite elbow. Lie down again.
Version 4: touch one knee with the elbow on the same side. Lie down
Version 5: any combination of the above.

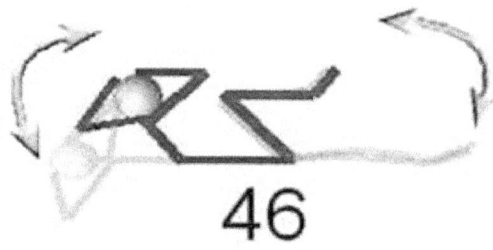

46

Exercise 46: Similar versions as in the exercise 45 but with both legs raised or lowered at the same time.

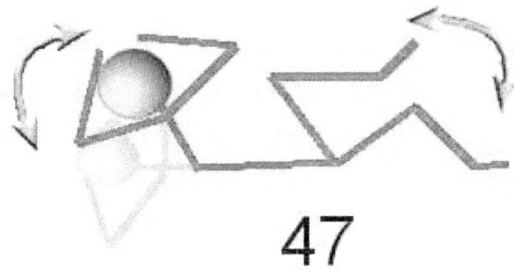

47

47. Similar versions as in the exercise 45, with knees kept bent all the time.

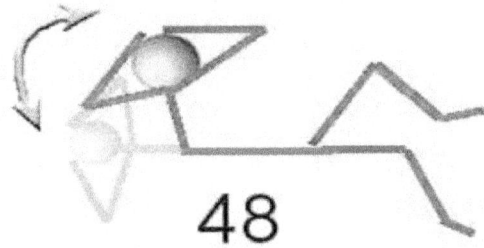

48

48. Raise your upper torso and twist the waist so that one elbow points towards the knee of the opposite side. Repeat the same on the other side. Keep the lower back straight all the time.

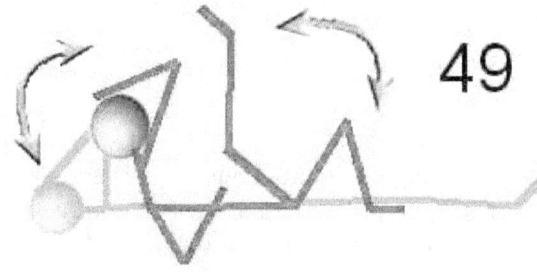

49

49. When raising one leg also raise the elbow on the opposite side, such that it touches the raised leg. Simultaneously, elevate the hip of the raised leg off the ground.

As the raising leg rounds the spinal curvature above lumbar regions, the grounded leg opposes that action. The balance between the two antagonistic actions requires your active participation is keeping straight and firm lower back. This reduces the risk of spinal injury.

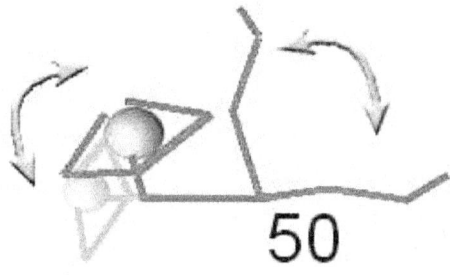

50

50. In this exercise, you have to use your hips extensively.

Version 1: Raise one leg as shown in the figure and point the elbow of the opposite side towards the raised knee. Simultaneously, elevate the hip of the raised leg off the ground and then return both the raised leg and scapula to the ground. Alternate sides in subsequent sets.

Version 2: The same like version 1 but with elbow-to-knee pointing occurring on the same side of the body.

Version 3: make version 1 to be one set by changing legs.

Version 4: make version 2 to be one set by changing legs.

Version 5: any combination of the above.

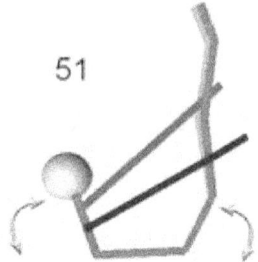

51

51. Raise both your upper torso and legs high in the air, with both hands pointing on one side. Simultaneously, elevate you hips off the ground. Maintain maximal contraction according to your need and then return only the hips and upper torso to the ground.

Make combinations with both sides either in one set or in different sets.

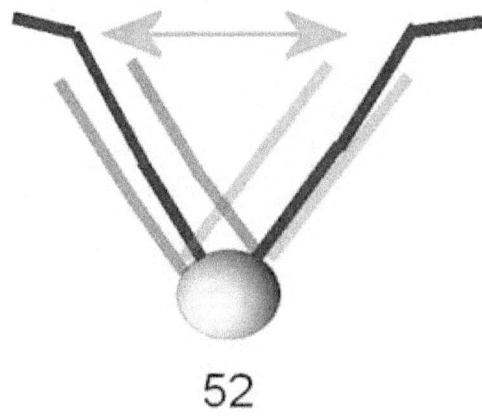

52

52. Starting position: lie down on your back with both straight legs raised high in the air and widely spread. The arms are extended right above your head.

Action: Raise your upper torso and reach one leg while elevating the hip at the side of the raised leg. Return the raised hip to the ground and your hands above your head. Repeat the same with the other leg. Do not forget to return your hands above your head and extended high in the air when not touching your legs.

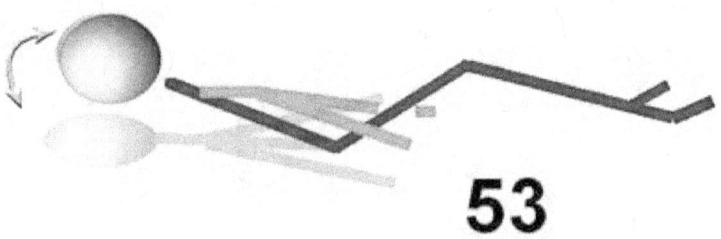

53

53. This is similar to the exercise 31 but here the heels are raised off the ground and the straight arms are moving in a horizontal plane, from left to right and visa versa, during maintaining balance. Remember to keep the lower back straight.

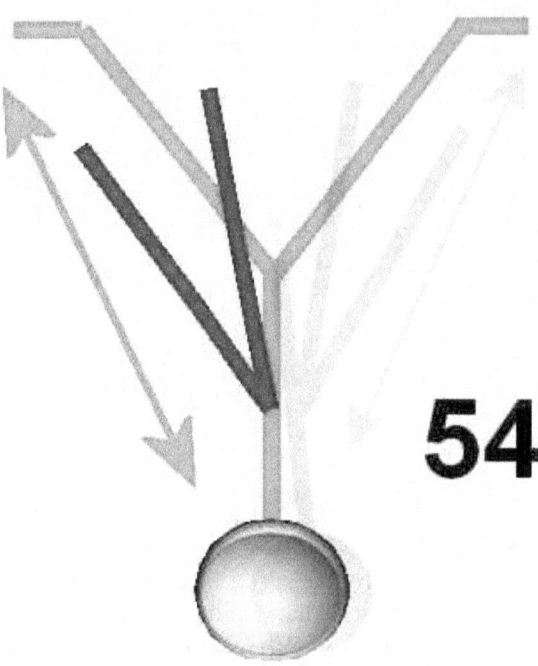

54. Lie on your back with legs spread wide apart. Raise your upper torso and touch one thigh with both hands. Lower your upper torso to the ground and then touch the other thigh with both hands.

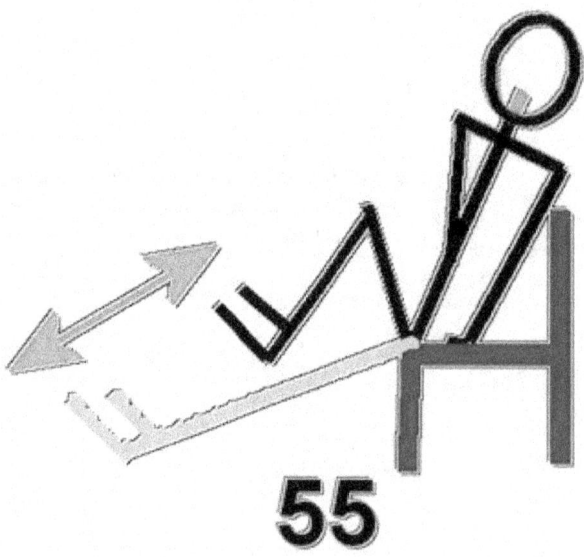

60

55. Either sit on a chair or on the ground. Support your body on your hands and balance your head such that you could pull your legs upwards.

Pull both flexed knees upwards on one side by keeping the lower back straight. After maximal hip flexion, extend both legs and then pull both knees on the other side.

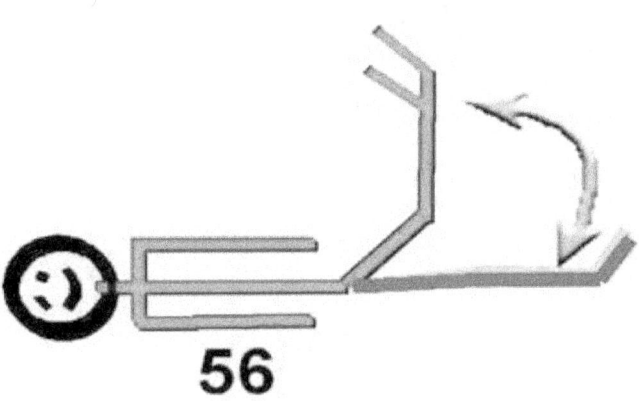

56. Raise your hips and keep the maximum contraction for a while. The arms are on the ground along your sides and help keeping your balance. Repeat by lowering and raising your hips off and onto the ground.

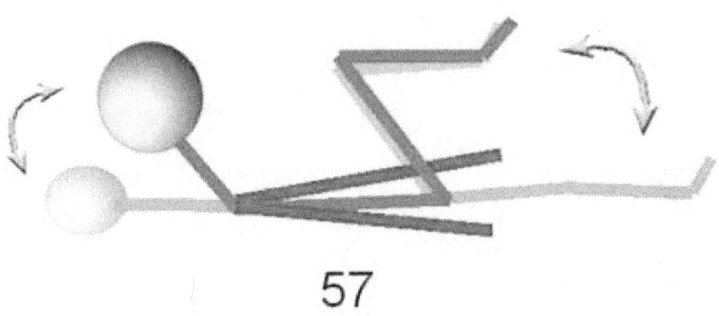

57. While keeping your hands under the buttocks, raise both upper torso and knees towards each other. Repeat by lowering and raising your upper torso and hips off and onto the ground.

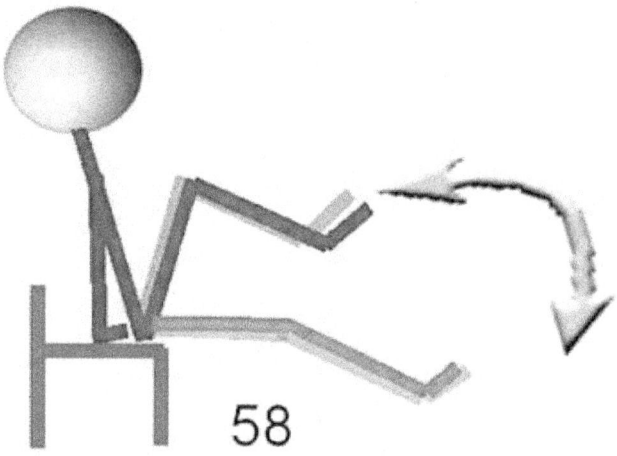

58. While sitting on the ground or on a chair raise the knees towards your head; keep the lower back straight. Maintain maximal contraction according to your needs.
Repeat by lowering and raising your hips and extending and flexing your knees.

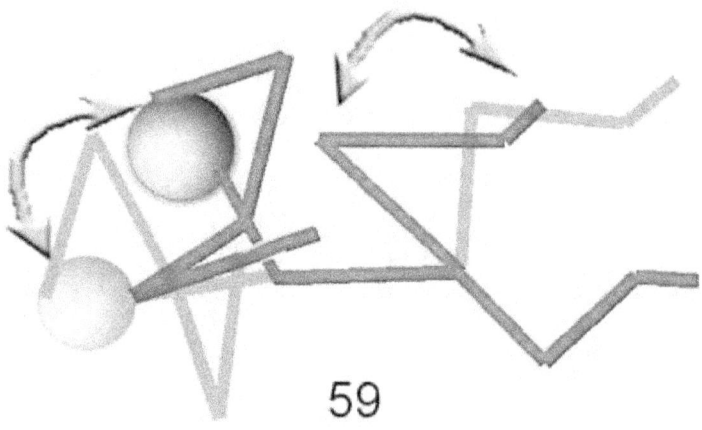

59. Lie down on one side with legs half-bent in the knees. Place the forearm on the lower side over your stomach and place the other hand behind your neck. Raise both the leg and the upper torso, on the upper side, so that the elbow touches the knee.
Alternate sides in subsequent sets.

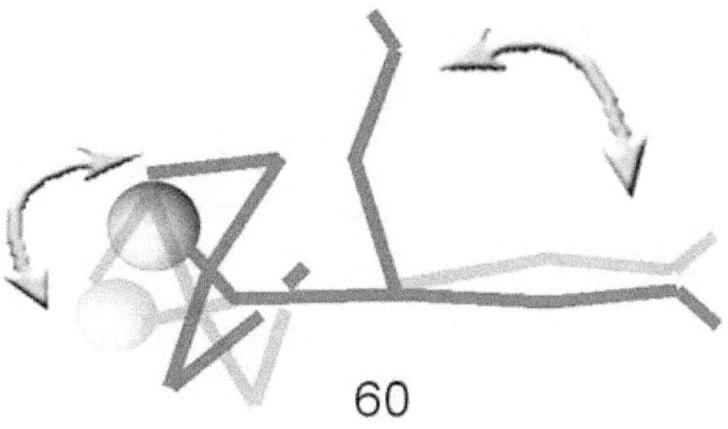

60

60. Similar to the exercise 59 but with both legs straight.
The side posture and the elevation of the upper torso challenge the Obliquus abdominis. The leg motion is supposed to pull and push the upper border of the pelvis, towards and away from the ribs, such that it intensifies muscular resistance.

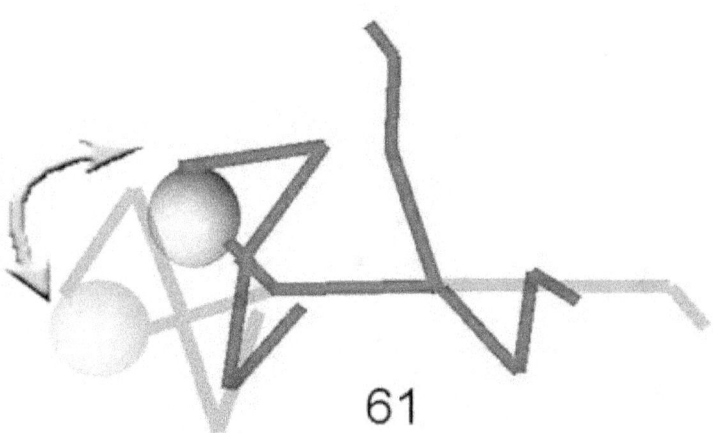

61

61. Similar to the exercise 59 but with variation in bending the knee of the raised leg. The raised leg is once lifted straight in the air and the other time bent in the knee. This exercise combines the exercises 59 and 60.

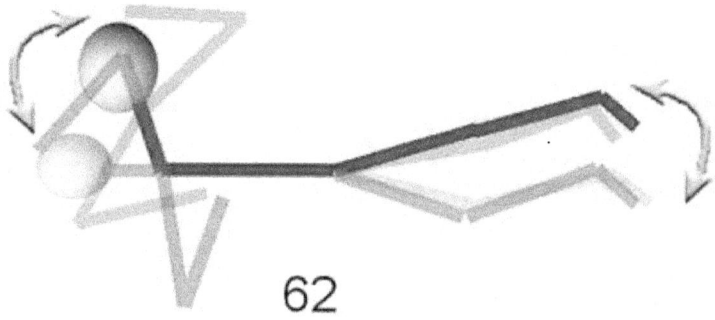

62

62. Lie down on one side with the lower forearm placed over your belly and the other hand placed behind your neck.
Raise both your upper torso and legs laterally. Repeat by lowering and raising the two to execute maximal contraction of the Obliquus abdominis muscle, followed by relaxation.

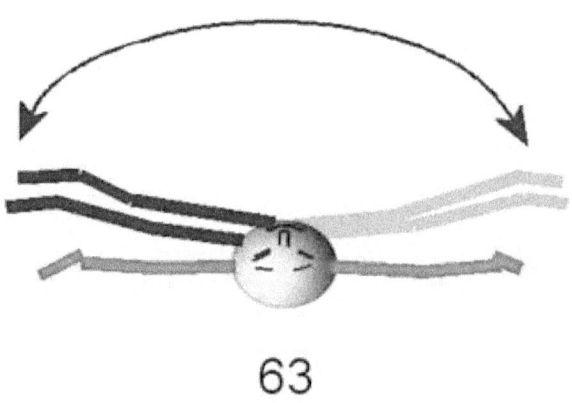

63

63. Lie down on your back with arms spread away from torso far to the sides. Raise your legs high in the air and twist your waist to one side until the legs touch both the ground and your extended hand on that side. Then return the legs high in the air again and twist your torso to the other side until the legs touch the ground and your extended hand on that side too.

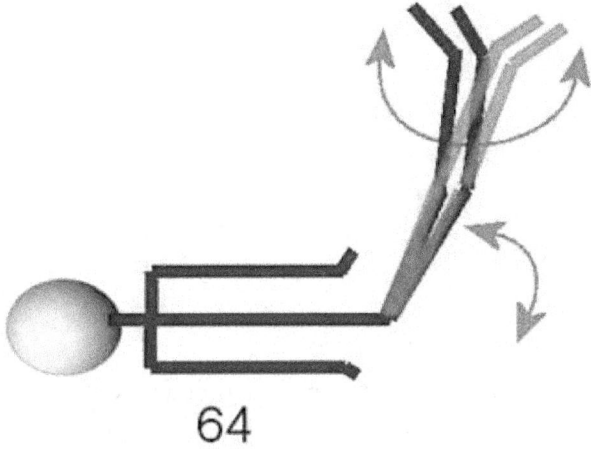

64

64. Lie down on your back with arms extended straight by your side and pushing against the ground supporting the balance.
Raise your hips in the air while twisting the waist on either side. This contracts the Rectus abdominis isometrically, while the Obliquus abdominis twist the pelvis over the spine.

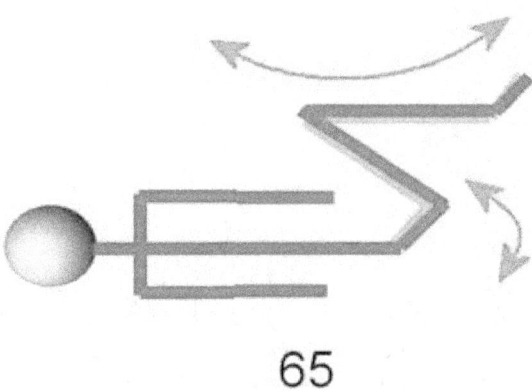

65

65. Starting position: lie down on your back with arms extended straight by your side and pushing against the ground supporting the balance. Both knees and hips are half-bent.
Ending position: make circles in the air with your knees. When the knees are closest to the chest the hips should be elevated from the ground.
Intensified version: elevate the hips off the ground and keep them above the ground all the time while making circles in the air with your knees.

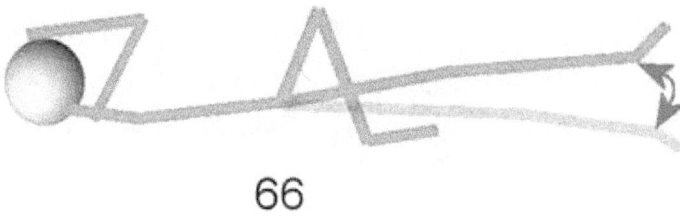

66

66. Lie down on your back with both hands behind your neck. Keep one leg fully stretched on the ground and the foot of the other leg just below the buttock on the same side. Action: raise your whole side of the extended leg above the ground by keeping your balance mostly on your left scapula and your grounded foot. After reaching maximal elevation and contraction, lower the elevated side of the body and then repeat the process. In this exercise, you use your grounded leg as an axis of rotation. Alternate sides between sets.

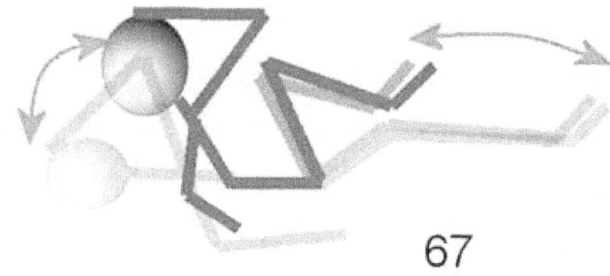

67

67. Lie down on one side and support yourself on the ground on your forearm. Place the hand of the higher arm behind your neck. Then raise your upper torso and your knees laterally so that your high elbow touches your knees. Preserve maximum contraction for a few moments. Repeat by lowering and then raising as described. Alternate sides between sets.

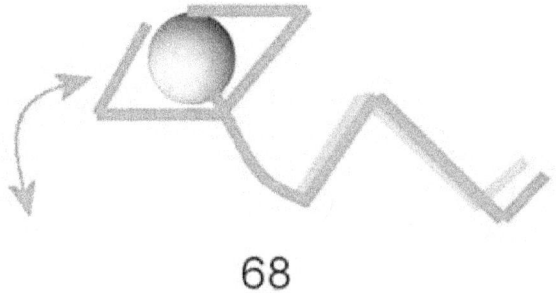

68

68. Lie down on your back with both hands behind your neck and legs half bent in knees and hips.
Raise upper torso towards your knees and simultaneously twist your waist to either side. After reaching maximal contraction, reverse the motion by lowering your upper torso. Repeat on the other side. Alternate sides between sets and in various combinations.

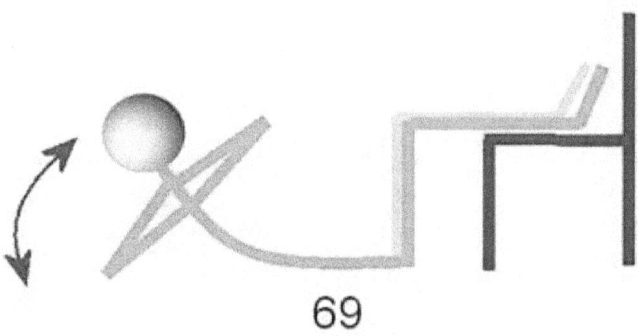

69

69. Lie down on your back with both hands behind your neck or on your chest. Anchor your legs under a fixed object such that the thighs assume perpendicular position to the ground.
Raise your upper torso and proceed by twisting your waist to either side. After reaching maximal contraction, reverse the motion and repeat the process on the same or opposite side of twisting.

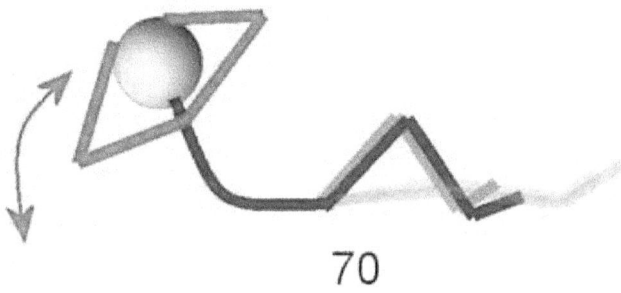

70

Lie down on your back with both hands behind your neck or on your chest
Drag your feet along the ground, half way to your pelvis, while raising your upper torso and twisting it to either side. After reaching maximal contraction, reverse the motion and repeat the process on the same or opposite side of twisting.

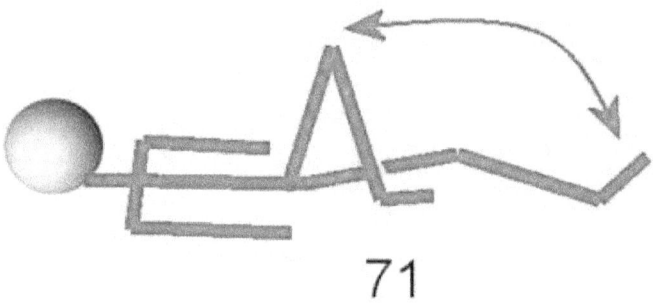

71

71. Lie down on your back with arms extended straight by your side and supporting the balance by pushing the arms against the ground. Slightly bend one leg at the knee, while the other leg is fully bent with the foot placed on the ground near pelvis.
The exercise consists of switching this leg arrangement by fully bending the slightly bent leg and extending the fully flexed leg. You should slide your heels over the ground during leg movement. Keep your abdomen tense and your lower back stuck to the ground all the time.

72

72. Lie down on your back with arms extended straight by your side and pushing against the ground supporting the balance.

Extend and raise one leg as shown in the figure, while the other leg is bent and its foot kept near the pelvis. Alternate this leg arrangement. Keep your abdomen tense and your lower back stuck to the ground all the time.

73

73. While kneeling, sit on the back of your heels. Keep the back straight, shoulders pulled backwards, hands on your thighs. Begin the exercise by exhaling deeply and slowly. Then, using the strength of your abdominal muscles, suck your stomach towards your spine and keep it that way for a while. Then take a breath, relax a little and repeat. Very good as a closing morning exercise.

6.3.2.2. Abdominal exercises on a suspension bar
These exercises make use of horizontal bar and are suitable for gym training.

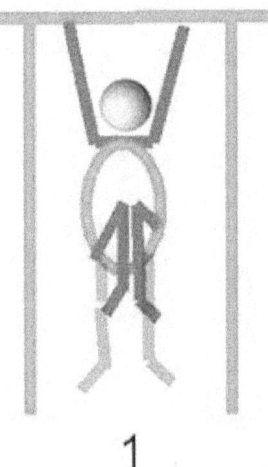

1

1. While hanging down from the suspension bar, keep your entire body falling free downwards.

Start the exercise by bending both knees and raising them towards the chest as high as possible. Maintain maximal contraction for a while before lowering your knees and reversing the motion. Repeat raising and lowering of the knees to make a set.

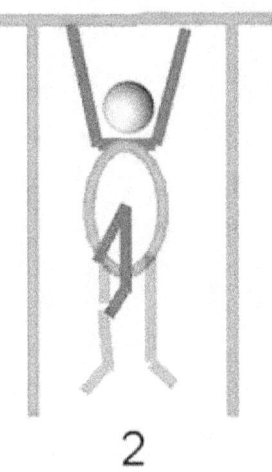

2

2. One leg is bent and the knee is brought as high as possible towards the chest, while the other leg hangs freely downwards. Legs can be used alternatively in one set, or a set can be designated for raising and lowering only one leg.

3

3. Like pedaling a bicycle, the legs are always bent in the knees. Raising and lowering continues, while the lower back is kept toned and the abdomen is sucked in.
The elevation of the scapulas and chest cage versus the vibration in the pelvic boundaries, due to leg motion, all stretch and challenge the abdominal muscles.

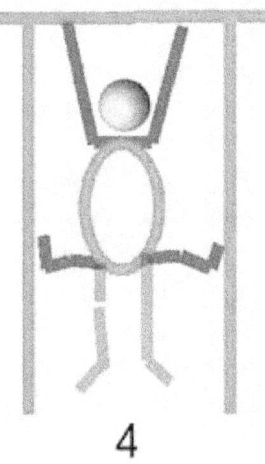

4. Both legs are kept straight, all the time. Start the exercise by flexing your hips in order to raise your legs from bottom to horizontal. Maintain the upper leg position for a while before lowering them down under controlled descent.

When both legs are raised, your body will bounce backwards due to the shift of the center of mass. This requires tensing your lower back in order to reduce the risk of injury.

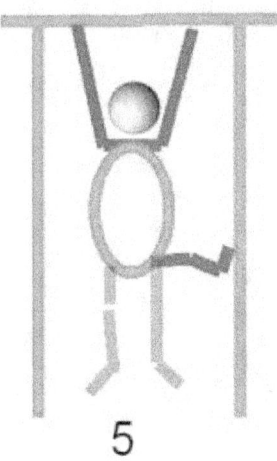

5. This is the same as exercise 2, yet the raising leg remains straight with only hip flexion.

Straight leg rising is a real challenge, since it stretches the hamstrings while requiring you to flex the Iliopsoas muscle.

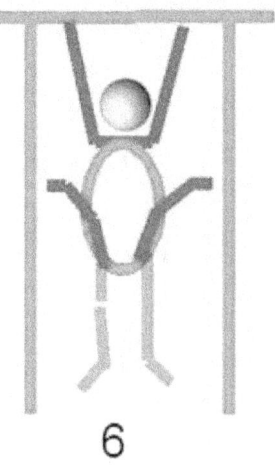

6. From the freely hanging position, start the exercise by elevating your knees towards your chin, with shins parallel to the ground.

Here you are flexing the hip joints and rounding the spine. Thus, both the Rectus abdominis and Iliopsoas muscles are engaged in isometric contraction.

7. From the freely hanging position, raise both straight legs to horizontal position by flexing the hip joints. Start the exercise by drawing figures of 8s, circles, zigzag pattern, or other with the tips of the feet of your straight legs.

This is a challenging exercise for the endurance and strength of both the Rectus abdominis and Iliopsoas muscles.

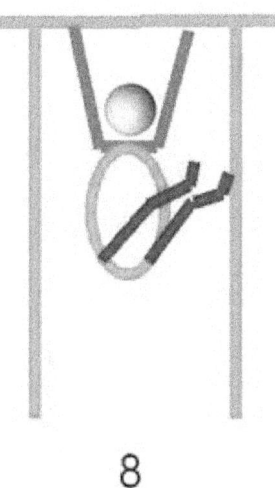

8

8. Start by your legs freely hanging downwards. Bend both hips and knees, while twisting your lower torso, in order to bring the knees together to one side, with thighs parallel to the ground.

Keep the maximum contraction for a while and then return the legs to hang freely downwards, before repeating on the other side.

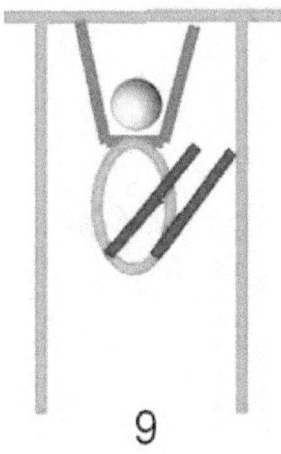

9

9. This is similar to the previous exercise but with the knees kept straight.

The compound actions of two hamstring muscles stretching, due to extended knees, are challenge to your ability to control the Rectus abdominis, Iliopsoas muscles, and Obliquus abdominis.

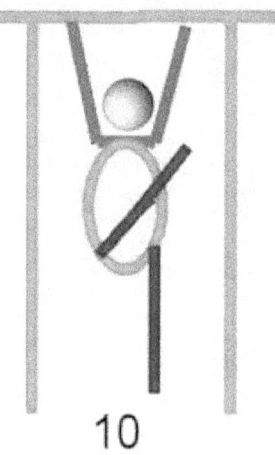

10

10. This is similar to the exercise 5, but with the straight leg shooting medially to the opposite side.

The raising leg creates two angular momenta: one that challenges the spinal erector and another that challenges the abdominal oblique muscles.

11

11. This exercise differs from exercise 2 in that the raising knee is both crossing the medial plane of the body (sagittal plane) and the hip is elevated by rounding the lower spine. This elevates one knee to the level of the shoulder on the opposite side of the knee.

12

12. This is an advancement of exercise 8 since the raising knees will reach the level of the shoulder on one side.

This exercise involves the Rectus abdominis, Obliquus abdominis, and Iliopsoas muscles. In addition, the Latissimus dorsi endure a lot of stretching during hanging from a suspension bar.

****

For most people, the best possible time for doing abdominal exercises is probably the early morning just after getting up from bed.

****

6.3.2.3. Abdominal exercises on a bench

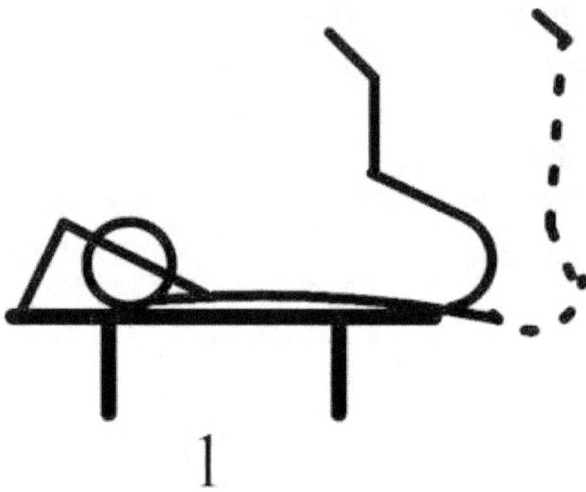

1. Starting position: lie back on a flat bench with your pelvis hanging over one edge while your hands are raised overheads and holding the other edge of the bench. Your legs should be kept slightly bent at the knees and the thighs perpendicular to the bench.
End position: pelvis is raised and the thighs are moved toward your stomach.

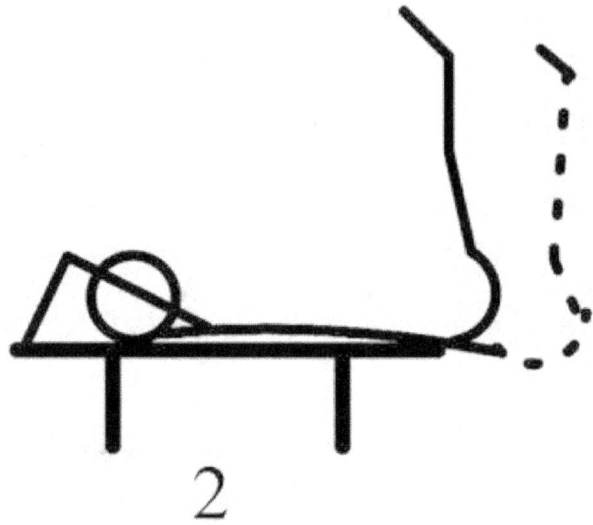

2. Similar to the exercise 1 but pelvis is elevated straight up towards the ceiling. This alters the relationship between the Abdominis rectus and Iliopsoas. The former would be contracting by shortening (isotonic), while the latter by elongating (eccentric).

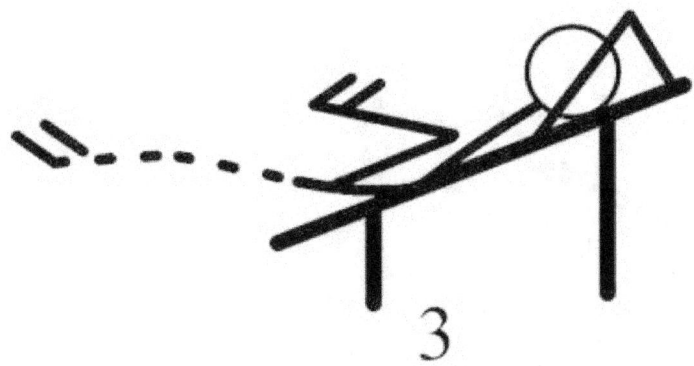

3. Lie on your back on an inclined bench with both legs kept parallel to the ground. Start by bending both knees and hips in order to bring the knees towards the chest.
The inclined bench adds constant isotonic tension to the hip and spinal flexors.

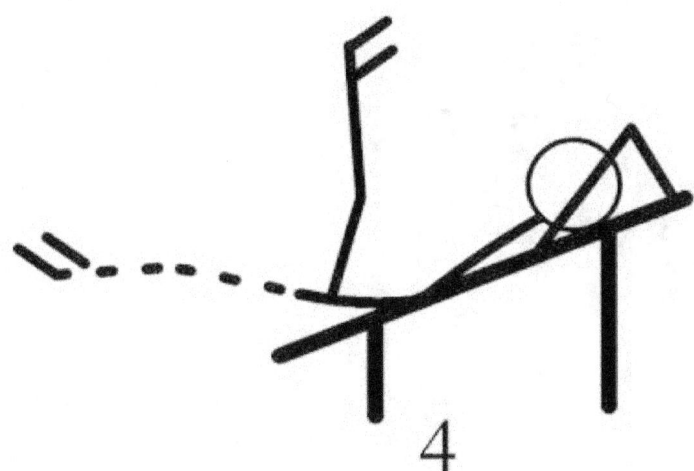

4. This differs from the previous exercise in that the knees are kept straight. When raising the legs, they end up in the perpendicular position to the ground. The extending legs increase the difficulty of resisting by adding more torque. Notice that the pelvis also departs the bench while the legs are being raised.

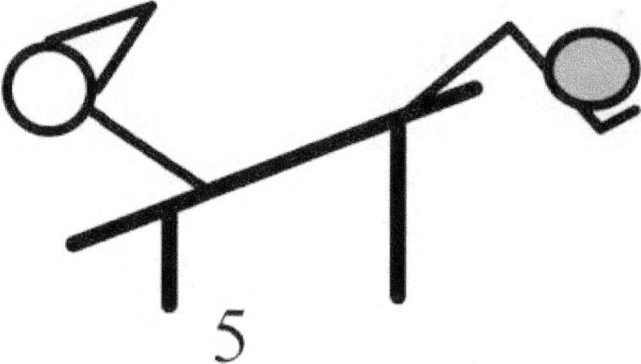

5. On a declined bench, raise just your upper torso with lower back stuck to the bench. The decline surface also adds an extra isometric tension to on the Rectus abdominis.

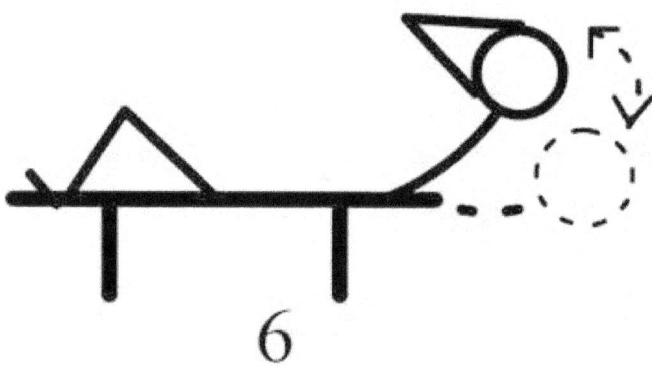

6. Lie on a flat bench with upper back over the edge. Raise your upper back by keeping your lower back stuck to the bench. The free suspension of the upper torso ensures constant isometric tension on the abdominal muscles.

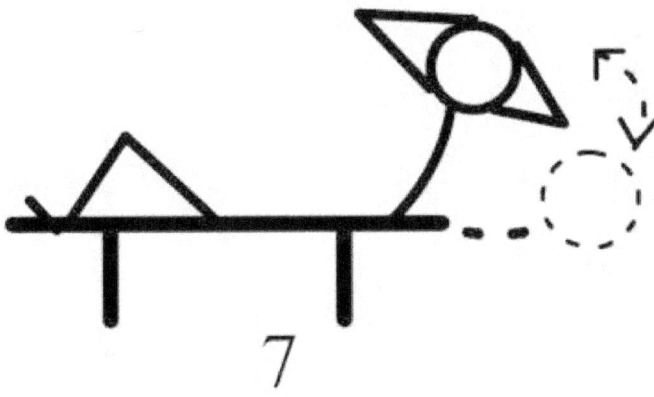

7

7. Similar to exercise 6 but twist the waist on either side when raising the upper torso. The twist activates the abdominal oblique muscles since it requires the ribs to be pulled anteriorly towards the symphysis pubis.

****

Abdominal exercises, as well as all other exercises, do not spot-reduce fat. Fat must be burned with long and consistent exercise and reduced by cutting down on caloric intake.

****